Visceral Leishmaniasis

Developments in Immunology

Visceral Leishmaniasis

Therapeutics and Vaccines

Awanish Kumar

ELSEVIER

ACADEMIC PRESS

An imprint of Elsevier

Academic Press is an imprint of Elsevier
125 London Wall, London EC2Y 5AS, United Kingdom
525 B Street, Suite 1650, San Diego, CA 92101, United States
50 Hampshire Street, 5th Floor, Cambridge, MA 02139, United States
The Boulevard, Langford Lane, Kidlington, Oxford OX5 1GB, United Kingdom

Library of Congress Cataloging-in-Publication Data
A catalog record for this book is available from the Library of Congress

British Library Cataloguing-in-Publication Data
A catalogue record for this book is available from the British Library

ISBN: 978-0-323-91124-5

For information on all Academic Press publications
visit our website at https://www.elsevier.com/books-and-journals

Publisher: Andre Gerhard Wolff
Acquisitions Editor: Linda Versteeg-buschman
Editorial Project Manager: Pat Gonzalez
Production Project Manager: Kiruthika Govindaraju
Cover Designer: Miles Hitchen

Typeset by SPi Global, India

Contents

Preface

This book mainly targets college students, scientists, and researchers interested or working in parasitology, applied microbiology, the pharmaceutical industry, medicinal chemistry, drug discovery, etc. Those from academia and industry involved in the aforementioned fields will find this book a valuable and fundamental guide. This compilation provides up-to-date information on the fatal disease visceral leishmaniasis (VL). This book provides a precise view of numerous components of VL that conveys the knowledge needed to understand and to improve the success in combating this disease. This includes Leishmania parasite culture, host immunity and immunomodulation, natural compounds effective against VL, animal models for VL, methodologies available for antileishmanial drug screening procedures against VL, current VL therapeutics, natural antileishmanial molecules, vaccine and vaccination-related information in VL, and amastigote-based vaccination strategy in leishmaniasis.

This book also provides concrete information on past and current difficulties in vaccine development in Leishmania. It reveals scientific opportunities and the potential impact of vaccines, including problems encountered in the development of VL vaccines. Future challenges for vaccines against VL are also discussed. Although vaccines are available against many diseases, there are still challenges ahead for a VL vaccine. The ideal vaccine must be safe and able to induce an immune response that is strong and effective for infectious diseases like VL, but a combination of chemotherapy (drug) and immunoprophylaxis (vaccine) could be ideal to win the battle against VL. Therefore, the author has tried to keep an appropriate balance between a general and an updated perspective of drugs and vaccines against VL. The author has also presented a state-of-the-art description of approaches/methodologies of antileishmanial screening and the status of vaccine development in VL. The topics discussed in this book collectively address the important material on this disease. The book provides consolidated information on VL (for educational purposes) and its future trends/possibilities (for research purposes).

Leishmaniasis: An overview

1

General account on leishmaniasis

Leishmaniasis is among the six most important tropical diseases according to the WHO [1]. Leishmaniasis is a group of tropical diseases caused by several species of protozoan parasites belonging to the genus *Leishmania* [2]. This ailment affects around 12 million people in 80 countries; it is estimated that there are about 2–3 million new cases each year. About 350 million people worldwide are at risk of infection. In most cases, leishmaniasis is transmitted zoonotically; however, in those cases where an animal reservoir is not known, an anthroponotic transmission is suspected. The vectors for the disease have been characterized as the female flying insects of the genus *Phlebotomus* (Old World leishmaniasis) and *Lutzomyia* (New World leishmaniasis). Most of the species are pathogenic to both man and lower vertebrates; they have been differentiated by means of genetic, biochemical, and immunological studies, and grouped into different complexes [3]. In humans, *Leishmania* spp. causes a variety of clinical diseases according to the ability of the organism to proliferate in deep tissue (37°C) or close to the surface of the skin at lower temperatures (25°C). The various manifestations of the disease have been used by the WHO as the basis to classify leishmaniasis into four clinical forms: visceral, mucocutaneous, cutaneous diffuse or disseminated, and cutaneous. Certain species of the parasite have been associated with different clinical forms of the disease. The *Leishmania donovani* complex causes visceral leishmaniasis. The *Leishmania tropica* complex is known to induce cutaneous lesions in the Old World and the *Leishmania mexicana* complex causes cutaneous and cutaneous diffuse leishmaniasis in several countries of Latin America. During their biological cycle, parasites of the *Leishmania* genus exist in two forms that develop in different hosts: a flagellated extracellular form known as promastigote and an intracellular one designated as amastigote. The form that infects both man and other vertebrate hosts is the promastigote, which is 15–20 μm in length, has a flagellum in its front part, and lives in the digestive tract of the insect transmitter of the disease. The amastigote, the parasitic form, has a diameter of approximately 2.3 μm and a rudimentary flagellate; it is located inside the host's macrophages. The vertebrate host is infected with the promastigote form of the parasite as a result of a sting by the vector insect. After this, the promastigotes are quickly phagocytosed by the macrophages of the host. Inside them the promastigotes change to the amastigote form. The clinical manifestation of the disease is a consequence of the amastigotes inside the macrophages.

Among the various form of leishmaniasis, cutaneous leishmaniasis is the least severe form and is generally considered an autolimited infection. Mucocutaneous leishmaniasis often results in facial disfiguration due to erosion in the mucocutaneous sites of the mouth and nose. On the other hand, leishmaniasis cutaneous diffuse or disseminated is characterized by the formation of nodules, plates, or multiple lumps, especially around

Visceral Leishmaniasis. https://doi.org/10.1016/B978-0-323-91124-5.00012-6

Table 1.1 Taxonomic classification of *Leishmania* parasites pathogenic for humans.[d]

Taxon	Old World[b]				New World[c]			
Family	*Trypanosomatidae*							
Genus	*Leishmania*							
Subgenus	*Leishmania*				*Leishmania*		*Viannia*	
Complex	*L. major*	*L. tropica*	*L. aethiopica*	*L. donovani*	*L. donovani*	*L. mexicana*	*L. braziliensis*	*L. guyanen*
Species	*L. major*	*L. tropica*	*L. aethiopica*	*L. donovani* *L. infantum*	*L. chagasi*	*L. mexicana* *L. venezuelensis* *L. garnhami* *L. amazonensis* *L. pifanoi*	*L. braziliensis* *L. peruviana* *L. colombiensis* *L. lainsoni* *L. shawi* *L. naiffi*	*L. guyanen* *L. paname*
Hybrid species[d]							*L. braziliensis/L. panamensis* *L. braziliensis/L. guyanensis*	

[a] The classification scheme is based on biochemical and molecular characteristics.
[b] Animal parasites of the Old World (*L. arabica, L. turanica,* and *L. gerbilli*) have not been listed.
[c] Animal parasites of the New World (*L. aristedesi, L. entrietti, L.forattinii, L. deanei, L. equatorensis,* and *L. bertigi*) are not shown; although human infection with this genera has not yet been identified, there is often a time lag between the discovery of new animal genera and the confirmation of human infection. In many cases, the relevant vectors have been found to bite humans as well as the normal animal hosts.
[d] Strains with phenotypic and genotypic characteristics of two species have been recorded in different geographic areas of Latin America; either they represent strains that originated from a common ancestor or they are the result of genetic exchange.

the face and on the external surface of the arms and legs. The taxonomic classification of *Leishmania* parasites is given in Table 1.1, which is adopted from [4].

Key facts about human leishmaniasis

Aetiological agent

Obligate intracellular protozoa of the genus Leishmania (order Kinetoplastida); about 21 of 30 species that infect mammals infect human beings.

Stages of parasite

Promastigote: It is flagellated form of *Leishmania* parasite found in sandflies having 15–20 mm long and 3.0–3.5 mm in breadth with 15–28 mm flagellum.

Amastigote: It is nonflagellated tissue form (2–4 mm in diameter); replicates in macrophage phagosomes in mammalian hosts.

Modes of transmission

Vector-borne–from the bite of infected female phlebotomine sandflies (about 2–3 mm long), which become infected by taking blood meal from the infected mammalian host. A total of about 30 species in the Phlebotomus genus (Old World) and Lutzomyia genus (New World) have been identified as vectors. Sandflies are relatively weak, noiseless fliers; they rest in dark, moist places, and are typically most active in evening and night-time hours.

Other modes–congenital and parenteral (i.e., by blood transfusion, needle sharing, and laboratory accidents).

Geographical distribution and estimates of numbers of cases

Leishmaniasis-endemic countries–88 countries. Leishmaniasis is endemic from northern Argentina to southern Texas (not in Uruguay, Chile, or Canada); in southern Europe; Asia (not southeast Asia); the Middle East; and Africa (particularly east and north Africa, with sporadic cases elsewhere), but not in Australia or Oceania.

Estimated number of people at risk of infection–about 350 million.

Estimated annual number of new cases of visceral leishmaniasis–about 500,000; more than 90% of worldwide cases are in Bangladesh, northeastern India (particularly Bihar State), Nepal, and Sudan (Old World), and in northeastern Brazil (New World).

Estimated annual number of new cases of cutaneous leishmaniasis–about 1·5 million; more than 90% of worldwide cases are in Afghanistan, Algeria, Iran, Iraq, Saudi Arabia, and Syria (Old World), and in Brazil and Peru (New World). The geographical distribution of cases evaluated in the developed world reflects travel and immigration patterns.

Clinical manifestations

Visceral leishmaniasis encompasses a broad range of manifestations of infection. Infection remains asymptomatic or subclinical in many cases, or can follow an acute, subacute, or chronic course. The classic kala azar syndrome is exemplified by patients such as those who found in Sudan and they are heavily infected throughout the mononuclear phagocyte system. They then develop a life-threatening disease after an incubation period of weeks to months. They suffer from fever and severe cachexia, hepato-splenomegaly (splenomegaly usually predominates), pancytopenia (anemia, thrombocytopenia, and leucopenia, with neutropenia, marked eosinopenia, and a relative lymphocytosis and monocytosis), and hypergammaglobulinemia (mainly IgG from polyclonal B-cell activation) with hypoalbuminemia.

Classification of leishmaniasis

Leishmaniasis occurs in more than 100 countries, from warm climates through subtropical to tropical climates. Southeast Asia and Australasia are the only large areas with suitable climates where the diseases are absent. On the basis of clinical manifestation, leishmaniasis is classified as:

Cutaneous leishmaniasis (CL)

It is frequently self-healing in the Old World, but when the lesions are multiple and disabling with disfiguring scars, it creates a lifelong aesthetic stigma. Its most severe form, recidivans leishmaniasis, is very difficult to treat, long-lasting, destructive, and disfiguring. Ninety percent of CL cases occur in seven countries: Afghanistan, Algeria, Brazil, Iran, Peru, Saudi Arabia, and Syria.

Diffuse cutaneous leishmaniasis (DCL)

This occurs in individuals with defective cell-mediated immune response. Its severity is due to disseminated lesions that resemble those of lepromatous leprosy, which never heal spontaneously and are subject to relapse after treatment with any of the currently available drugs. Because of the devastating consequences to the patient, it is recognized as a special public health problem. It is a chronic, progressive, polyparasitic variant that develops in the context of leishmanial-specific anergy and is manifested by disseminated nonulcerative skin lesions, which can resemble lesions of lepromatous leprosy. It is caused by *L. aethiopica* (Old World) and *L. mexicana* species complex (New World).

Mucocutaneous leishmaniasis (MCL)

Also known as "espundia," it causes the extensive destruction of oral-nasal and pharayngeal cavities with hideous disfiguring lesions, mutilation of the face, and great suffering for life.

Visceral leishmaniasis (VL)

Also known as "kala azar," VL is the most severe form (nearly always fatal if left untreated), characterized by undulating fever, weight loss, splenomagaly, hepatomegaly and/or lymphadenopathies, and anemia. It causes large-scale epidemics with high fatality rates. After recovery, patients may develop a chronic CL form called "post-Kala azar dermal leishmaniasis," (PKDL) which usually requires a long and expensive treatment.

Post-kala azar dermal leishmaniasis (PKDL) is a syndrome that develops at variable times after the resolution of visceral leishmaniasis. It can be associated with a relapse of the visceral disease, and is manifested by skin lesions that can be of various types and initially are most prominent on the face. People with chronic PKDL can serve as reservoir hosts of infection.

Viscerotropic leishmaniasis is an oligoparasitic syndrome with nonspecific manifestations caused by visceral infection with *L tropica*, which more typically is dermotropic. Parasitologically confirmed cases (*L tropica* confirmed as the aetiological agent in eight) were noted among US servicemen in the Persian Gulf conflict in the 1990s.

Visceral leishmaniasis is concentrated in eastern Africa, particularly Sudan and Kenya, and on the Indian subcontinent in Bangladesh, northeast India, and Nepal. Post-kala azar dermal leishmaniasis has a similar distribution, though the proportion of treated kala azar cases that develop this condition varies. Infantile visceral leishmaniasis is characteristically circum-Mediterranean in its distribution, extending east through southwest Asia to China, and west to Central and South America, where most cases occur in northeast Brazil. This last form is increasingly rare in its classical infantile form, but has expanded to affect adults who are infected with the HIV virus or who are otherwise immunocompromised (www.who.org).

Among all forms of Leishmaniasis, VL is the most severe clinical form of the disease and it can be fatal when not treated. It is characterized by its effect on the internal organs, particularly the liver, spleen, and bone marrow. Focused information on VL is discussed in the next section.

An outline of fatal VL

VL or kala azar was first described as a fatal disease in humans by two scientists, Leishman and Donovan, in 1903. *Leishmania donovani* (a protozoan parasite) is the causative agent of VL. This does not mean that VL did not exist before 1903. As far back as 1870 in Saudi Arabia, Cunningham recorded a similar disease in Saudi Arabia that occurred in 1885 (caused by a protozoan parasite) and was named later as *Leishmania tropica* (a causative agent of cuteneous leishmaniasis; CL). Nicolle in 1908 reported that mammals other than humans, including dogs, could act as reservoir hosts for the *Leishmania* parasite. Using human volunteers, Swaminatha et al. in 1942 proved that the *Leishmania* parasite could be transmitted by *Phlebotomus* sandflies (www.who.org). Different species of sandfly transmit *Leishmania* and their reservoirs include canine, wild rodents, and human. Female phlebotomine sandflies become infected by taking blood meal from an infected mammalian host. Sandflies are relatively weak, noiseless fliers; they rest in dark, moist places, and are typically most active in evening and night-time hours.

Within the insect host, *Leishmania* is present as a flagellated promastigote form in the sandfly vector invertebrate host. Upon infecting the mammalian host, it differentiates into the smaller aflagellated round amastigote stage. Amastigotes are obligate intracellular parasites of macrophages (and rarely other cell types), where they survive and multiply within a phagolysosome compartment. After an infected blood meal, the vector takes about 4–18 days, according to the species of leishmania, to be able to transmit the infection to the victim.

Leishmaniasis is one of the major infectious diseases affecting the poorest regions of the world. There are two million new cases each year in the 88 countries where this disease is endemic, and 80% people earn less than $2 a day. There are three major types of Leishmaniasis that range from self-limiting CL (cutaneous leishmaniasis) to fatal VL if not treated successfully. VL is difficult to treat and there is increasing resistance developing against the currently available antileishmanial drugs [5]. New foci of VL are identified every year in different parts of the world. This may be due to the emerging resistance of sandflies toward insecticides and the resistance of the parasite to the existing chemotherapy [6]. In developing and underdeveloped parts of the world, acquired immunosuppressive syndromes (including AIDS) also increase the risk of VL [7]. Based on these observations, there is clearly an urgent need for effective antileishmanial or vaccine development against fatal kala azar because it is the most dreaded one and is the major cause of mortality and morbidity.

Burden of VL

The trypanosomatid parasite of the genus leishmania is the etiological agent of a variety of disease manifestations collectively known as leishmaniasis. VL is the most devastating form of leishmaniasis. VL is fatal and its major clinical symptoms include prolonged fever, anemia, splenomegaly, prominent wasting, and death from organ failure and opportunistic infections when left untreated [6]. More than 90% of global VL cases are reported from Bangladesh, Brazil, India, Ethiopia, South Sudan, and Sudan. An estimated 50,000–90,000 new cases of VL occur worldwide annually and more than 95% of new cases reported to WHO occurred in 10 countries: Brazil, China, Ethiopia, India, Iraq, Kenya, Nepal, Somalia, South Sudan, and Sudan. According to the WHO, 20,000–40,000 deaths occur every year from leishmaniasis (https://www.who.int/leishmaniasis/en/). Deaths are reported from poor countries; hence, researchers are linking this disease to poverty, economic underdevelopment, and various environmental changes such as deforestation, urbanization, migration of people into endemic areas, building of dams, etc. [8], (www.who.int/whr/2002/whr2002_annex3.pdf).

VL occurs more in developing countries, especially in poorer sections of society. VL has been present on the Indian subcontinent for more than 100 years. In the Bihar province of India, it was first reported in the Purnea district as Kala Dukh (disease) about 100 years ago. Prior to that, the disease was prevalent mostly in Assam and Bengal. VL is caused by the bite of the infected female sandfly Phlebotomus argentipes, which is the vector. Man-sandfly-man is the VL cycle. At present, the disease is present in India in almost all districts of Bihar, four districts of Jharkhand, five districts of Uttar Pradesh, and 10 districts of West Bengal. VL has spread from Bihar to West Bengal and Bangladesh in the east, Uttar Pradesh in the west,and Nepal in the north. Now, several cases have been reported from other states of India (Tamilnadu and Orissa) to a lesser extent. In VL, extreme emaciation develops, the skin becomes dark, and hair becomes dry. Certain complications such as dysentery, pulmonary tuberculosis, herpes zoster, and epistaxis may occur. The main drug treatments recommended for VL were first introduced more than 80 years ago. However, that is changing, and some new therapies are becoming available.

There are many species of Leishmania that cause leishmaniasis and many species of sandfly also exist as carriers. The epidemiology is extremely diverse and >20 Leishmania species are pathogenic for humans. More than 30 sandfly species are proven to be vectors of the Leishmania parasite. There are two main epidemiological entities: (1) zoonotic, which includes animal reservoir hosts in the transmission cycle, and (2) anthroponotic, in which man is the sole source of infection for the vector. In terms of control, priority has been given to anthroponotic foci as the feasibility for control is the highest (well-known previous successes) and these foci are the source of severe and deadly VL epidemics. In the early 1990s in southern Sudan, MSF Holland reported 100,000 deaths among a population of less than 1000,000. The majority (90%) of VL cases occur in the poor rural and suburban areas of five countries: Bangladesh, India, Nepal, Sudan, and Brazil.

A great deal of attention has been focused on VL during the past decade because of etiological and behavioral factors and the resistance of the parasite to drugs. This attention is also largely because of the socioeconomic conditions of the population at risk and because the untreated disease is always fatal.

The overall prevalence of leishmaniasis cases is 12 million people and the population at risk is 350 million [9]. However, it is very difficult to provide realistic estimates of those infected versus those at risk. There is probably an even greater difference between the number of cases actually occurring and the number reported, due to several factors such as the distribution of transmission sites within endemic areas. It was estimated that in India about 1.2 million people were infected with Leishmania, resulting in 1000 deaths per year. Most of them belong to the poor class.

VL is the serious form of leishmaniasis. It is a life-threatening disease involving the whole reticuloendothelial system, including the bone marrow and skin, where drug delivery becomes difficult. In this disease, the macrophages, which are normally destined to kill any invading or foreign organism, become the site not only for lodging the parasites but also for their multiplication. This results in the depression of the immune status of the host, which drastically alters the immune system of naturally dying as well as drug filled parasites to further aggravate the toxic condition leading to derangement of various physiological functions. These conditions favor a suitable situation for intercurrent infections to further worsen the situation. The management of such conditions becomes more complex with existing drug therapy. With the available diagnostic methods, the disease diagnosis will be made at a later stage. Also, monitoring the drug effect is not possible. Hence, this research project has been selected to contribute to the better management of kala azar by creating an efficacious, safe, inexpensive, and short-course oral treatment from ayurvedic medicine for developing and underdeveloped countries.

The development of a potent drug against VL is a challenging problem, as the parasites are well protected within the host cells. In addition, due to massive immunosuppression, there is a lack of appropriate host response against the disease [10–12]. Consequently the patient fails to respond satisfactorily to the available antileishmanial drugs. For any compound or agent to be effective, it is necessary that it must reach the parasite in a desired concentration after overcoming the barriers imposed by the host cells. At the same time, the drug should have no harmful effect on the host cells [13, 14]. Besides having antileishmanial efficacy, compounds/ayurvedic drugs that are also capable of restoring immune function will present an additional advantage. To address these current challenges, much research is being done in India and abroad. Current challenges in this field include the emergence of leishmaniasis in new geographic areas, host populations, and persons infected with HIV; also the increasing number of treatment failures in India; and the geographic distribution of these areas. These challenges have led us to consider more carefully the emergence of drug-resistant parasites as the main cause of sodium antimony gluconate (SAG)/sodium stibogluconate (SSG) unresponsiveness in Bihar, India. Now, we need applicable, rapid diagnostic tests for the field as well as efficacious, safe, inexpensive, and short-course oral treatment regimens. It is evident from a literature review that natural products possess antileishmanial activity. If successfully developed into a drug, this will be helpful toward better management of the disease because these products in general show low/tolerable toxicity and can be administered orally. In this regard, we selected ayurvedic drug extracts to screen out antileishmanial activity. For screening of medicinal plants/compounds, various techniques have been developed and tried to assess their antileishmanial action [13, 15, 16].

Although leishmaniasis is not a household name like malaria, the disease caused by infection with *Leishmania* continues to have a major impact on much of the world's population. The disease is endemic throughout parts of Africa, India, the Middle East, Southern Europe, and Central and South America. With the advent of the human immunodeficiency virus (HIV) epidemic, leishmaniasis has surged as a reactivating infection in AIDS patients in many parts of the world. The six different species of *Leishmania* recognized to cause disease in humans are very similar morphologically but produce strikingly different pathological responses. The only feature common to all is the chronicity of disease manifestations. The infection may be predominantly visceral, as in VL or Indian kala azar; restricted to the skin, as with the chronic ulcer of the Oriental sore; or spread to the mucous membranes to produce the disfiguring South American espundia. In the vertebrate host, *Leishmania* parasites survive and multiply intracellularly in mononuclear phagocytes as nonmotile amastigotes about 2–4 μm in diameter. Transmission to the vertebrate host is through *Phlebotomus* sandflies, in which parasites develop and replicate as 20 μm flagellated promastigotes. In the fly, the parasites undergo a developmental program starting with the amastigote in the blood meal, continuing through several stages of promastigote maturation, and culminating with the infectious metacyclic form. The environmental cues, which trigger this program, are not well understood, but temperature and pH appear to play a role. Leishmaniasis is considered a zoonosis, and humans are generally accidental hosts. An important exception to the zoonotic character of leishmaniasis is that the reservoir for the cutaneous disease caused by *Leishmania tropica* in the Middle East and the visceral disease in India is probably made up of oft-infected humans. The animal reservoir shows geographic variation and includes rodents, dogs, and other mammals.

VL is endemic in northeast India, especially in Bihar, West Bengal, Assam, and certain pockets of eastern Uttar Pradesh. Estimates indicate that approximately 2,50,000 new cases of VL develop each year in Bihar alone. The factors determining the susceptibility or resistance to VL remain unclear, but the genetics of the host may play a major role. The incubation period of the disease may be as short as 2 weeks or as long as 9 years. In diseased individuals, VL is characterized by intermittent fever, massive hepatosplenomegaly, anemia, thrombocytopenia, and polyclonal B-cell activation with hypergammaglobulinemia. VL is often fatal. A notable feature of clinical VL is the apparent cellular energy to parasite antigens.

VL coinfection with HIV

VL (occasionally other syndromes) is emerging as an important opportunistic infection among people with HIV-1 infection [17]. In fact, the parasite may be a cofactor in the pathogenesis of HIV infection; a major surface molecule, the lipophosphoglycan, of L. *donovani* induces the transcription of HIV in CD4 cells. To date, leishmania and HIV coinfection has been reported from 31 countries with most of the cases from southern Europe, where 25%–70% of adult patients with VL are coinfected with HIV, and 1.5%–9.0% of patients with AIDS develop leishmaniasis. The dual trends of the encroachment of leishmaniasis on urban areas, and of HIV infection on rural areas, are likely to lead to more coinfected patients elsewhere. VL occurs in many countries having warm climates through subtropical to tropical climates. Southeast Asia and Australasia are the only large areas with suitable climates where the diseases are absent. VL is concentrated in eastern Africa, particularly Sudan and Kenya, and on the Indian subcontinent, in Bangladesh, northeast India, and Nepal. Post-kala azar dermal leishmaniasis has a similar distribution, though the proportion of treated kala azar cases that develop this condition varies. Infantile VL is characteristically circum-Mediterranean in its distribution, extending east through south-west Asia to China, and west, to Central and South America where most cases occur in northeast Brazil. This last form is increasingly rare in its classical infantile form but has expanded to affect adults who are infected with the HIV virus, or who are otherwise immunocompromised.

The existing antileishmanial drugs such as pentamidine and amphotericin-B are potentially toxic. Their success rates are poor, especially in immune-depressed individuals, most notably AIDs patients. Further, treatment failure with antileishmanial drugs has also been reported worldwide, and these have often been interpreted as drug-resistant cases. Thus, the intervention of an effective and reliable drug is urgently needed. After explaining the maintenance of the leishmania parasite and infection in the next section, a few chapters describe the immunomodulation and immunomodulatory activity of natural compounds in VL.

The surveillance data for southern Europe are as follows: 1461 cases of coinfection reported for January 1990 through June 1998 (835 [57.2%] from Spain); and 717 cases for January 1996 through June 1998 (412 [57.5%] from Spain) [18] (P Desjeux, personal communication).

Chemotherapy for VL

The control of VL remains a major global problem. The well-documented control measures include mainly four strategies: drugs, vaccines, hygiene, and health education. Under the existing national scenario of illiteracy, poverty, population increases, and pollution, the implementation of the last two strategies seems impractical. Further, the presence of any vaccine or therapeutics are the mainstay of remedy and disease cure. But there are no reliable and safe drugs available for VL. Even today, the drugs used to treat leishmaniasis are more than 50 years old. The following groups of compounds have been used as antileishmanial drugs (Fig. 1.1):

- Antimony compounds–trivalent (antimony potassium tartrate, tartyart emetic) and pentavalent antimonials (sodium stibogluconate, meglumine antimoniate).
- Aromatic diamidine–stibamine, hydroxystibamine, and pentamidine.
- Antibiotics–liposomal amphotericin B.
- New drugs–miltefosine, allopurinol, aminosidine, paramomycin.
- Immunotherapy–IFN-γ + sodium stibogluconate.

Historically, the chemotherapy of leishmaniasis has been based on the use of toxic heavy metals, particularly antimony compounds. When this kind of treatment is not effective, other medications used include pentamidine and amphotericin B. All these pharmaceuticals require administration by injection and clinical supervision or hospitalization during treatment due to the severity of the possible secondary effects. The leishmanicidal agents with the most favorable therapeutic index are the antimony compounds known as antimonials: sodium stibugluconate (pentosam) and meglumine antimonate (glucantime). However, the antimonials cause serious side effects such as pain at the site of injection, stiff joints, gastrointestinal problems,

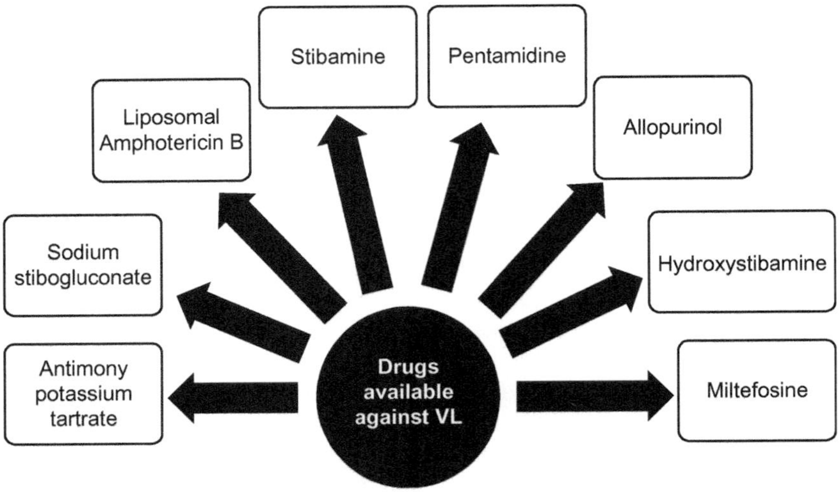

Fig. 1.1 Drugs available for VL chemotherapy.

cardiotoxicity and, in some cases, hepatic and renal insufficiency. These products also require lengthy treatments and their cost is high.

Pentamidine is an aromatic derivative of diamidine; it is toxic to a number of protozoa and useful for the treatment of VL cases that do not respond to antimonials. Among the side effects associated with the intravenous administration of pentamidine are hypotension, tachycardia, fainting spells, upset stomach, and severe headaches, in addition to skin eruptions, abnormal function of the liver, hypoglycemia, and renal disfunction. Amphotericin B is a polyene antifungal agent that represents an adequate and different treatment to the antimonials. This pharmaceutical product is also associated with an important number of side effects, including the alteration of the renal function in approximately 80% of treated individuals. Although there are a number of products being developed, to date none of them has been demonstrated to be fully effective against *Leishmania* parasites. These pharmaceutical products include allopurinol, primaquine, compound WR6062, and miltefosine, but they are yet to be marketed or are in the process of being marketed.

Among antimony compounds, pentavalent antimonials (less toxic in comparison to trivalent antimonials) are still the drug of choice against VL, despite their side effects such as anorexia, vomiting, nausea, and malaise. They are also reported to have adverse effects on the heart and pancreas [19]. Failure in prolonged treatment with antimonials has also been reported worldwide, and these have been interpreted as drug-resistant cases [20, 21]. Second-line drugs such as pentamidine are more effective than antimonials, but they also induce serious disorders in the glucose metabolism and nephrotoxicity [22]. The drug amphotericin B and its derivatives are used successfully, though their applicability in the field is limited due to toxicity, long-term therapy, and complicated mode of administration. Liposomal amphotericin B is expensive and toxic. Miltefosine, a new low-cost oral drug with tolerable toxicity, is undergoing phase III trials for Indian VL.

SAG unresposiveness and resistance

VL treatment has revolved around sodium stibogluconate (SbV) for more than six decades. VL is considered one of the most neglected tropical diseases. As there is no vaccine and as the impact of vector control is limited, chemotherapy is the key strategy in VL control and cure. No effective vaccines are available against leishmaniasis, and the treatment relies on chemotherapy [23–25]. The first-line drugs against all forms of Leishmania infection consist of pentavalent antimony Sb (V)-containing drugs such as pentostam and glucantime. Resistance to this drug class has been described in several parts of the world, but has reached epidemic proportions in Bihar, India [26, 27]. This resistance to Sb (V) drugs is one of the WHO's antimicrobial resistance priorities (www.who.int/infectious-disease-report/2000).

The state of Bihar carries the largest share of the world's burden of antimony-resistant VL. Unresponsiveness to antimony has developed steadily to such an extent that antimony must now be replaced, despite attempts to stop this resistance progression by increasing the dose and duration of therapy. The classic second-line treatments are unsuited: pentamidine is toxic and its efficacy has also declined, and amphotericin B deoxycholate is effective but requires hospitalization for long periods while toxicity is common. Liposomal amphotericin B is very effective and safe but currently unaffordable because of its high price. Miltefosine–the first oral drug for visceral leishmaniasis–is now registered and marketed in India and is effective, but should be used under supervision to prevent misuse. Paromomycin (or aminosidine) is effective and safe, and although not yet available, a regulatory submission is due soon. To preserve the limited armamentarium of drugs to treat VL, drugs should not be deployed unprotected; combinations can make drugs last longer, improve treatment, and reduce costs to households and health systems.

The situation is particularly grave in Bihar, India, known as the "heartland of kala azar." Here, the burden of disease has increased steadily [28]; the disease is spreading, and unresponsiveness to antimonials has severely compromised disease control [26, 27]. Currently, of all regions, Bihar is facing the most immediate public health problem, with a lack of suitable treatment options [21]. Current alternative treatments are amphotericin B and its lipid formulations, pentamidine, miltefosine, and paromomycin [29,30].

For years, Bihar has been facing the problem of untreatable VL. Standard first-line antimony has become progressively inadequate. We consider that a central element to this continuous erosion of efficacy has been the use of subtherapeutic doses of antimony because of a combination of policy decisions, substandard drug quality, inadequate prescribing, and poor compliance, compounded by epidemiological features favoring resistance. Although the extent and gravity of the problem are quite unique to antimony in Bihar today, we may witness this same phenomenon where similar conditions occur.

Antimony unresponsiveness in mucocutaneous and visceral leishmaniasis is a serious clinical problem. Information on the mechanisms and characteristics of drug resistance in parasites that suggest chemotherapeutic strategies to overcome resistance is of practical importance. New formulations, therapeutic switching of the established drugs amphotericin B and paromomycin, and the serendipitous discovery of miltefosine have markedly improved leishmaniasis chemotherapy in the past several years. With the availability of complete genome sequences for all three pathogens, and methods for rapid validation of targets, it is hoped that much-needed amelioration will occur soon. More than 60% of patients with VL in Bihar do not respond to treatment with pentavalent antimonials. This is now considered to be due to acquired resistance. Although this class of drugs has been used for more than 60 years for leishmaniasis treatment, it is only in the past 2 years that the mechanisms of action and resistance have been identified, related to drug metabolism, thiol metabolism, and drug efflux. With the introduction of new therapies, including miltefosine in 2002 and paromomycin in 2005–2006, it is essential that there be a strategy to prevent the emergence of resistance to new drugs. Combination therapy, monitoring of therapy, and improved diagnostics could play an essential role in this strategy.

The standard treatment of human VL involves the use of pentavalent antimony (SbV) compounds. In recent years, increasing numbers of clinical failures of treatment with SbV have been reported, probably due to the development of parasite resistance to this compound. The mode of action and mechanisms of resistance to SbV have not been fully elucidated [31].

The threat of drug resistance is a major problem for all infectious diseases. Although the selection of resistant lines of Leishmania has long been a part of laboratory studies on antileishmanial drugs, it is only in the past decade that acquired resistance has become a clinical threat, and this mainly is restricted to pentavalent antimonial drugs in one focus in India. Studies over the past decade have demonstrated an ever-decreasing response rate of VL cases to antimonial treatment in Bihar. In this major focus of VL, more than 60% cases fail to respond to pentavalent antimonials [21]. After more than 70 years of use, the antileishmanial mechanism of action of pentavalent antimonials is still not clearly understood. This makes an adequate explanation of the mechanism(s) of resistance not possible. However, recent studies have described five areas of parasite and host biochemistry that are involved in antimonial sensitivity. First, the lack of sensitivity of promastigotes and the sensitivity of amastigotes to pentavalent antimonials has been shown to be related to the inability of promastigotes to metabolize the pentavalent compound to the trivalent compound, whereas this metabolism occurs in amastigotes [32, 33]. At present we do not know whether "resistant" amastigotes are also deficient in their ability to reduce SbV to SbIII. It has been suggested that the resistance property selected in promastigotes is present in amastigotes transformed from these clones; in vivo L. *panamensis* amastigotes derived from an SbIII-resistant promastigote line were more resistant to SbV treatment than the wild type [34]. Thiols in host cells and parasites also play a role. Thiols of host cell macrophages reduce SbV to the more active SbIII [32]. Parasite thiols (trypanothione, glutathione) are considered to have a critical role in the detoxification of heavy metals in trypanosomatids. Studies have shown that trivalent antimony and arsenic-resistant Leishmania parasites have increased levels of thiols (trypanothione, glutathione, ovothiol). These thiol levels have been related to the amplification of enzymes involved in the synthesis of glutathione (−glutamylcysteine synthetase, gsh1) and polyamines (ornithine decarboxylase), the two precursor molecules of trypanothione in some studies [35] but not in others [36]. Another potential part of antimony detoxification is the conjugation of metal with a thiol, a process mediated by glutathione-s- transferase (GST). Increased GST expression and levels have recently been reported in antimony-resistant isolates with concomitant increased levels of spermidine and glutathione [37]. A trypanothione S- tranferase, if present, could play a role here as well. Finally, levels of Sb in cells can be controlled through uptake or efflux. The accumulation of SbV and SbIII into promastigotes and amastigotes has been shown to be by different transport systems [38]. Although Sb accumulation was lower in resistant than sensitive forms, levels of accumulation could not be correlated to sensitivity in wild-type cells. The efflux of the thiol-Sb conjugate is also of importance with increased levels of efflux in some cases [36], or sequestration in an intracellular vacuole.

The above discussion indicates the lack of acceptable, safe, and potentially effective antileishmanial drugs so far. A vaccine against VL as also not been developed yet. Nevertheless, there are at least a half-dozen vaccine candidates in the pipeline, but none of them have entered the market as approved vaccines. On this condition and background, the need for either novel antileishmania agents or effective vaccines is clear. Therefore, the authors have compiled herein the information, progress, and status of both antileishmanial drugs and vaccines for the ease and better understanding of researchers and readers in this area, especially under various chapters of this book.

References

[1] A.L. Wilson, O. Courtenay, L.A. Kelly-Hope, T.W. Scott, W. Takken, S.J. Torr, S.W. Lindsay, The importance of vector control for the control and elimination of vector-borne diseases, PLoS Negl. Trop. Dis. 14 (1) (2020), e0007831.

[2] S. Pandey, D. Dhami, A. Jha, G. Shah, A. Kumar, M. Samant, Identification of trans-2-cis-8-matricaria-ester from the essential oil of Erigeron multiradiatus and evaluation of its antileishmanial potential by in-vitro and in-silico approaches, ACS Omega 4 (11) (2019) 14640–14649.

[3] A. Kumar, Transmission of Leishmaniasis from human to other vertebrate: A rapid zooanthroponotic evolution, Int. Microbiol. 22 (3) (2019) 399–401.

[4] S.L. Garcia, Diagnostic medical parasitology, fourth edition, 2001, pp. 205–209. Washington, D.C. USA.

[5] R.N. Davidson, Visceral leishmaniasis in clinical practice, J. Infect. 2 (1999) 112–116.

[6] A. Kumar, S.C. Pandey, M. Samant, Slow pace of antileishmanial drug development, Parasitol Open 4 (e4) (2018) 1–11.

[7] D. Wolday, N. Berhe, H. Akuffo, S. Britton, *Leishmania*-HIV interaction: immunopathogenic mechanisms, Parasitol. Today 15 (1999) 182–187.

[8] The World Health Report. Geneva: WHO, 2002: 192–7. www.who.int/whr/2002/whr2002_annex3.pdf.

[9] L.H.C. Makala, B. Baban, Novel therapeutic approaches to Leishmania infection, in: D.M. Claborn (Ed.), Leishmaniasis – Trends in Epidemiology, Diagnosis and Treatment, IntechOpen, 2014, https://doi.org/10.5772/58167.

[10] J. Alexander, S.L. Croft, Chemotherapy and immune response, in: D.T. Hart (Ed.), Leishmaniasis: The Current Status and New Strategies for Control, Nato-ASI Series A, Life Sciences, 1988, p. 163.

[11] A. Bertoli, A.V. Greoo, Visceral leishmaniasis presenting with hypertrigycerdaemia, Lancet 11 (1982) 504.

[12] J.K. Nag, M.P. Ghosh, Kala-azar with pulmonary tuberculosis, J. Indian Med. Assoc. 25 (4) (1955) 138.

[13] J.W. Mikhail, N.S. Mansour, M.J. Khayyal, Leishmania donovani. : Therapeutic action of antimony dextron glycosides (R.L. : 712) in the golden hamster, Exp. Parasitol. 37 (1975) 348.

[14] P.C. Sen Gupta, A. Chatterjee, A.M. Mukherjee, Antimony dextron in the treatment of leishmaniasis, J. Indian Med. Assoc. 37 (12) (1961) 585.

[15] W.L. Hanson, W.L. Chapman Jr., W.L. and Kinnamon, K.E., Testing of drugs for antileishmanial activity in golden hamsters infected with L. donovani, Int. J. Parasitol. 7 (1977) 443.

[16] L.A. Stauber, E.M. Franchino, J. Grun, An eight day method for screening of compounds against Leishmania donovani in the golden hamster, J. Protozool. 5 (1958) 269.

[17] J. Alvar, P. Aparicio, A. Aseffa, et al., The relationship between leishmaniasis and AIDS: the second 10 years, Clin. Microbiol. Rev. 21 (2) (2008) 334–359.

[18] Centers for Disease Control and Prevention, Sexually Transmitted Disease Surveillance 2018, U.S. Department of Health and Human Services, Atlanta, 2019, https://doi.org/10.15620/cdc.79370.

[19] C.P. Thakur, K. Kumar, Efficacy of prolonged therapy with stibogluconate in kala-azar (including post kala-azar dermal leishmaniasis), Trans. R. Soc. Trop. Med. Hyg. 78 (1990) 391.

[20] E.A.G. Khalil, A.M. El-Hassan, E.E. Zijlstra, F.A. Hashim, M.E. Ibrahim, H.W. Ghalib, M.S. Ali, Treatment of visceral leishmaniasis with sodium stibogluconate in Sudan: Management of those who do not respond, Ann. Trop. Med. Parasitol. 92 (2) (1998) 151.

[21] S. Sundar, Drug resistance in Indian visceral leishmaniasis, Trop. Med. Int. Health 6 (2001) 849–854.

[22] T.K. Jha, Evaluation of Diamidine compound (pentamidine isothionate) in the treatment of resistant cases of kala-azar occurring in North Bihar, India. Trans. R. Soc. Trop. Med. Hyg. 77 (1983) 163.

[23] B.L. Herwaldt, Leishmaniasis, Lancet. 354 (9185) (1999) 1191–1199, https://doi.org/10.1016/S0140-6736(98)10178-2.

[24] J.D. Berman, Human Leishmaniasis: clinical, diagnostic, and chemotherapeutic developments in the last 10 years, Clin. Infect. Dis. 24 (1997) 684–703.

[25] H.W. Murray, Clinical and experimental advances in treatment of visceral leishmaniasis, Antimicrob. Agents Chemother. 45 (2001) 2185–2197.

[26] S. Sundar, D.K. More, M.K. Singh, V.P. Singh, S. Sharma, A. Makharia, P.C. Kumar, H. W. Murray, Failure of pentavalent antimony in visceral leishmaniasis in India: report from the center of the Indian epidemic, Clin. Infect. Dis. 31 (2000) 1104–1107.

[27] S. Sundar, D.K. More, M.K. Singh, et al., Failure of pentavalent antimony in visceral leishmaniasis in India: report from the center of the Indian epidemic, Clin. Infect. Dis. 31 (2000) 1104–1107.

[28] C.P. Thakur, Socio-economics of visceral leishmaniasis in Bihar (India), Trans. R. Soc. Trop. Med. Hyg. 94 (2000) 156–157.
[29] R.N. Davidson, Practical guide for the treatment of leishmaniasis, Drugs 56 (1998) 1009–1018.
[30] P.J. Guerin, P. Olliaro, S. Sundar, et al., Visceral leishmaniasis: current status of control, diagnosis, and treatment, and a proposed research and development agenda, Lancet Infect. Dis. 2 (2002) 494–501.
[31] M. Ephros, A. Bitnun, P. Shaked, E. Waldman, D. Zilberstei, Stage-specific activity of pentavalent antimony against Leishmania donovani axenic amastigotes, Antimicrob. Agents Chemother. (1999) 278–282.
[32] C.D.S. Ferreira, P.S. Martins, et al., Thiol-induced reduction of antimony (V) into antimony (III): a comparative study with trypanothione, cysteinyl-glycine, cysteine and glutathione, Biometals 16 (3) (2003) 441–446.
[33] P. Shaked-Mishan, N. Ulrich, et al., Novel intracellular SbV reducing activity correlates with antimony susceptibility in Leishmania donovani, J. Biol. Chem. 276 (6) (2001) 3971–3976.
[34] B.L. Travi, Y. Osorio, et al., Selection for resistance to trivalent antimony SbIII in Leishmania viania panamensis promastigotes leads to resistance to pentavalent antimony SbV in intracellular amastigotes, in: 52nd Annual meeting of the American society of Tropical Medicine and Hygiene. Philadelphia, Penn, USA, 2003.
[35] A. Haimeur, C. Brochu, et al., Amplification of the ABC transporter gene PGPA and increased trypanothione levels in potassium antimonyl tartrate (SbIII) resistant Leishmania tarentole, Mol. Biochem. Parasitol. 108 (1) (2000) 131–135.
[36] F.E. Arana, J.M. Perez-Victoria, et al., Involvement of thiol metabolism in resistance to glucantime in Leishmania tropica, Biochem. Pharmacol. 56 (9) (1998) 1201–1208.
[37] M. Chatterjee, M. Case, et al., in: Identification of a glutathione S- transferase in antimony resistant Leishmania donovani promastigotes, XIV Molecular Parasitology meeting, Woods Hole, USA, 2003.
[38] C. Brochu, J. Wang, et al., Antimony uptake systems in the protozoan parasite Leishmania and accumulation differences in antimony resistant parasites, Antimicrob. Agents Chemother. 47 (10) (2003) 3073–3079.

Maintenance of *Leishmania* parasite

2

In vitro maintenance of *Leishmania* parasite

Culture media required: Culture media are prepared to culture the *Leishmania* parasite [1–3]. Below are some common culture media for Leishmania promastigote cultivation.

Rosewell Park Memorial Institute (RPMI)–1640 Medium

The medium was prepared by adding 13.4 g RPMI–1640 (Sigma, United States) powder supplemented with 0.3 g/L L-glutamine in 500 mL of triple distilled water (TDW) followed by the addition of 2 g $NaHCO_3$, 2 g glucose, 5.94 g N-2-hydroxy-ethyl-piperazine-N-2-ethane sulphonic acid (HEPES) with constant stirring for about 45 min; the pH was adjusted to 7.2 ± 0.2. Finally, the volume was adjusted up to 1000 mL with TDW. The medium was filtered using Millipore (pore size 0.2 μm), and 40 mg gentamycin and 10% sterile FBS were added. The filtered medium was tested for sterility by putting a drop in a thioglycollate medium (2.98 g). The thioglycollate medium was dissolved in 100 mL TDW (pH 7.2 ± 0.2) and was sterilized by autoclaving at 15 lbs for 15 min. It was incubated at 37°C for 24 h. Because thioglycollate supports rich bacterial growth, even slight bacterial contamination grows luxuriously and makes a white ring on the top of the medium. Sterile RPMI-1640 was finally stored at 4°C.

Leibovitz (L-15) medium

L-15 medium (14.7 g; Sigma, United States) supplemented with 0.3 g L-glutamine was dissolved in 800 mL of distilled water followed by the addition of 2 g glucose with constant stirring for 30 min. About 2.95 g of tryptose phosphate broth (TPB) (HiMedia Labs, India) was dissolved in 100 mL of TDW and autoclaved at 15 lbs pressure for 15 min. In the above-mentioned L-15 medium, 10% TPB and 10% FBS (Sigma, United States) were added and the pH was adjusted to 7.2 ± 0.2. Finally, the volume was adjusted 1000 ml with triple distilled water. The mixture was passed through Millipore filters (0.2 μm); 40 mg gentamycin was added and stored at 4°C.

Medium 199 (M199)

This medium was prepared by adding 13.4 g M199 (Sigma, United States) powder in 500 mL of deionized water, followed by the addition of $NaHCO_3$ (2.0 g), glucose (2.0 g), and HEPES (5.94 g) with constant stirring for about 30 min. The final pH

Visceral Leishmaniasis. https://doi.org/10.1016/B978-0-323-91124-5.00011-4

was adjusted to 7.2 ± 0.2. Finally, the volume was adjusted up to 1000 mL. The medium was filtered through a 0.2 μm Millex-GV filter unit (Millipore, United States). Gentamycin (40 mg/L) and 10% heat-inactivated FBS (HiFBS) were added and the medium finally stored at 4°C.

Dulbecco's modified eagle's medium (DMEM)

DMEM was prepared by adding 13.4 g of powdered DMEM (Sigma, United States) in 500 mL of deionized water, followed by the addition of $NaHCO_3$ (2.0 g), glucose (2.0 g), and HEPES (5.94 g) with constant stirring for about 30 min. The final pH was adjusted to 7.2 ± 0.2. Finally, the volume was adjusted up to 1000 ml. The medium was filtered through 0.2 μm Millex-GV filter units (Millipore, United States). Gentamycin (40 mg/L) and 10% HiFBS were added and the medium finally stored at 4°C.

***Leishmania* strains cultured in vitro:** A number of isolates were procured from VL patients in different parts of the world. They were cultivated in the above-discussed medium. After incubation, the culture was ready to harvest for further study.

L. donovani *WHO reference strain Dd8 (MHOM/IN/80/Dd8)*

This strain was procured in 1981 from Imperial College, London, by Dr. P.C.C. Garnham. This strain was originally isolated from a human patient from Bihar, India, in 1979. It was found to be virulent in animals and produced fatal disease. Thus, it was considered appropriate to use this as the test parasite for various experimental studies. The promastigote form of the parasite was cultured at 26°C in an RPMI-1640 medium (Sigma, United States) supplemented with 10% HiFBS, penicillin (100 U/mL), and streptomycin (100 μg/mL). Parasites were subcultured every third or fourth day.

Clinical isolates of L. donovani

Several field isolates were procured from patients from endemic regions and put into cultures. The designation of responsive patients was based on the absence of fever, clinical improvement with a reduction in spleen size, and the absence of parasites in the splenic aspirates. The biopsy materials were cultivated initially at 26°C in NNN-agar tubes. Subsequently, promastigotes were cultured in cM199 supplemented with 20% HiFBS, 100 U/mL penicillin, and 100 μg/mL streptomycin at 26°C in 25 cm^2 tissue culture flasks [2].

Experimental host for in vivo maintenance of *Leishmania*

The leishmania parasites are maintained in vivo in hamsters through amastigote-to-amastigote passages or at times by inoculating the stationary phase promastigotes. The hamster (experimental host) should be small for manipulative handling, of standard breed with minimum genetic variation, and should be easily available in adequate

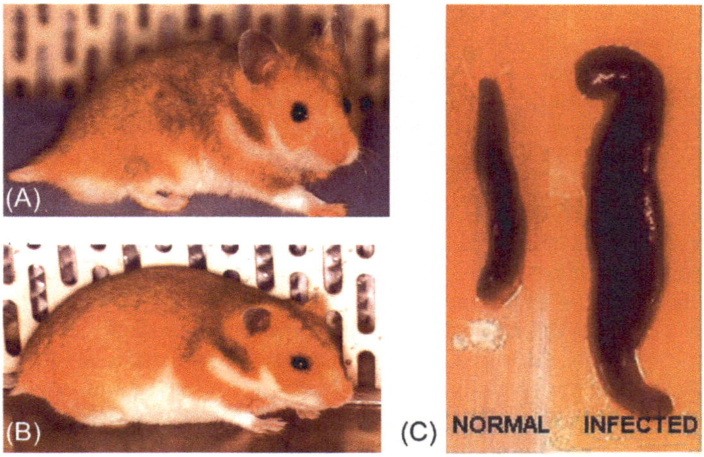

Fig. 2.1 Golden hamster as experimental host: (A) Naive hamster, (B) L. *donovani*-infected hamster, and (C) Spleen of normal and L. *donovani*-infected hamsters.

numbers. Syrian golden hamsters (*Mesocricetus auratus*) satisfy almost all the conditions mentioned above. Thus, the L. *donovani* hamster system appears appropriate as an experimental model for the investigation of VL. Therefore, hamsters are employed as the experimental host for *Leishmania* infection. Laboratory-bred male golden hamsters weighing 45–50 g are generally used as the experimental host (Fig. 2.1). To maintain the hamsters, they are housed in plastic cages in climate-controlled rooms and fed with standard rodent food pellets (Lipton Ltd., Bombay) and water *ad libitum*. Experiments on the animals should be performed according to the guidelines of the Institutional Animal Ethics Committee (IAEC).

In vivo maintenance of L. *donovani*

(i) *Preparation of infective inoculum*

By Promastigotes:

Promastigotes harvested from 7 to 10-day-old cultures were washed 2–3 times by centrifugation at 2000 rpm for 20 min (4°C). The pellet was suspended in a phosphate buffer saline (PBS) and the promastigote number was adjusted to 1×10^8/0.1 mL.

By Amastigotes:

The hamsters with well-established infections (15–60 days old) were sacrificed. The spleen was removed aseptically in a sterile Lock's solution (4 g NaCl, 0.10 g KCl, 0.10 g $CaCl_2$, 0.15 g KH_2PO_4, 1.250 g glucose was dissolved in 500 ml distilled water, pH 7.2) and cut into small pieces. The splenic tissue was homogenized gently with the help of a motor-driven tissue homogenizer consisting of a glass mortar and a Teflon pestle at 4°C. The suspension was centrifuged at 500 rpm for 10 min at 4°C, allowing the tissue debris to settle down. The supernatant was centrifuged at 2000 rpm for 20 min at 4°C. The supernatant was discarded and the sediment containing amastigotes was resuspended in PBS, adjusting the count with the help of a Neubaur's chamber (hemocytometer) to the concentration of 1×10^7 amastigotes per 0.1 mL of the cell suspension.

(ii) *Inoculation to animals*

Male golden hamsters weighing 40–45 g were inoculated intracardially with 1×10^7 amastigotes or 1×10^8 promastigotes [4]. The animals were kept in plastic cages (38 × 27 × 13 cm) with a diet of standard rodent pellets (Lipton India Ltd.) and water *ad libitum*.

(iii) *Assessment of infection*

The infection levels in the infected hamsters were assessed on day 20–25 postinfection by performing splenic biopsies. Hamsters were anesthetized by injecting 80 mg/kg of sodium pentabarbitol (Thiopentone) through intraperitoneal (i.p.) route. An incision of about 1 cm over the gastrosplenic region was made aseptically and the omentum was drawn exposing the spleen. A small piece of spleen was cut and kept on a filter paper. The spleen was carefully inserted back to its position and the muscles and skin sutured using a curved needle (3/4″, 22–25 gauge) and braided nylon stitch thread (0 No.). Tincture benzoin was layered on the sutured portion and neosporin powder was sprinkled on it. One day prior to operation and 2 days postoperation, neomycin sulfate (10 mg/kg) was administered orally to prevent bacterial infection. The biopsied splenic tissue was dabbed over filter paper so as to absorb excess blood and then impression smears were made on clean microscopic slides. The smears were air dried, fixed in absolute methanol for 60 s, and stained with 10% Giemsa for 45 min. The slides were washed with running tap water, then air dried and examined for the number of amastigotes per 1000 macrophage nuclei under the 100 × oil-immersion objective of a light microscope.

In vivo:

Parasitic assessment in experimental hamsters was carried out by splenic biopsy as described in earlier charper Chapter 3. The splenic dab smears were fixed in methanol, stained with 10% Geimsa, and examined for amastigotes under a microscope. The results were interpreted as the number of amastigotes/1000 cell nuclei of macrophages and were expressed as described in earlier charper Chapter 5.

In vitro:

The presence of parasites in the spleen was simultaneously confirmed by an in vitro culture. A small piece of spleen was cultured in NNN-medium vials, with RPMI-1640 as an overlay as described in earlier charper Chapter 3. After 7–10 days of incubation at 26°C ± 10°C, the vials were checked for the growth of promastigotes.

Primary culture of promastigotes

Hamsters infected for 45–60 days were dissected to isolate the infected spleen. The infected spleen was given 2–3 washings in 1 × PBS containing gentamycin. The splenic tissue was then transferred in an incomplete RPMI medium containing a gentamycin antibiotic at a concentration of 2 mg/100 mL media. After washings, the tissue was transferred into iRPMI (5–10 mL) containing an antibiotic. Flasks were examined daily for up to 10–20 days. When the presence of moving promastigotes was observed in the flasks, 20 μL of primary culture was transferred to fresh flasks containing 5–10 mL complete RPMI [5].

Macrophage cell line maintenance

J774, a mouse macrophage cell line, and THP-1, a nonadherent human macrophage cell line, were maintained in Dulbecco's modified eagle's medium (DMEM) and the RPMI-1640 medium, respectively, supplemented with 20% hiFBS, penicillin (100 U/mL), and streptomycin (100 μg/mL) at 37°C in 5% CO_2 in a humidified atmosphere. J774 cells were subcultured every second or third day, whereas THP-1 undifferentiated cells were subcultured twice a week. THP-1 was differentiated by incubation for 24 h in a medium containing 20 ng/mL phorbol myristate acetate (PMA, Sigma) before establishing the infection and carrying out drug susceptibility studies [6]. These cell lines are used for the maintenance of the amastigote form of the *Leishmania* parasite in different culture media.

Suitabitity of L. *donovani* as experimental parasite

The first and foremost prerequisite of the test parasite is that it should interact with the host's immune system and with chemotherapeutic agents in a manner that exists in clinical situations. A virulent strain of L. *donovani* (World Health Organization reference strain) was found virulent in animals and produced fatal disease. Thus, it was considered appropriate to use this as the test parasite for various experimental studies. The chemotherapeutic responses of existing standard antileishmanial drugs (stibanate, pentamidine, and amphotericin B) in experimental models were similar to those obtained in human cases [7,8]. During the course of infection, there is a reversal of the albumin/globulin ratio [9], excessive elevation of the antibody level, and severe depression of the cell-mediated immune (CMT) responses [10,11], besides severe anemia [12] and leucopenia [8]. Significant changes have also been observed in certain biochemical parameters such as the elevation of alanine aminotransferase and alkalinephosphisphatase, the depletion of the cholesterol level, and a reduction in total protein [9]. These parameters justify the usage of L. *donovani* (Dd8 strain) as the experimental parasite.

Cryopreservation of L. *donovani* promastigotes

To avoid the risk of losing parasites due to any adverse condition, they were stored in liquid nitrogen [13]. For cryopreservation, washed promastigotes were suspended in the RPMI-1640 medium supplemented with 10% FCS and low glycerol (glycerol 28 mL, sorbitol 72 mL of 4.2%, 0.9% NaCl) was added in equal volume. It allows storing the Leishmania promastigotes at −60°C for up to 18 months. A gradual decrease in temperature of 1–2°C/30 min was carried out before storage in liquid nitrogen. The revival of stored parasites was done by thawing the vials by rolling them between the palms. A gradual dilution of the sample with a culture medium was carried out after thawing, followed by centrifugation and resuspension in a fresh medium. The study is divided into three groups:

- In vitro study
- In vivo study
- Clinical trial will be done in few selected cases after confirming experimental results.

Cultivation of L. *donovani* parasite

(a) **In biphasic medium:** Prior to subculture, the NNN culture vials and RPMI-1640 were brought to room temperature. The overlay RPMI-1640, 2–4 mL, was added in each vial to which 1 mL of positive axenic culture containing $1–1.5 \times 10^6$ promastigotes from the stationary phase (Metacyclic form) was inoculated. These culture vials were incubated at 26°C ± 1°C. The parasite passes through four different phases during cultivation: (i) lag phase, (ii) logarithmic growth phase, (iii) stationary phase, and (iv) decline phase [14]. Subcultures of the parasites were carried out every week for continuous maintenance.

(b) **In monophasic medium:** For a bulk number of promastigotes, the parasites were cultivated in a monophasic medium (L-15). The axenic culture of promastigotes of approximately 0.1 million/mL was inoculated in a medium and incubated at 26°C ± 1°C. After 3–5 days, the culture was ready to harvest.

Assessment of L. *donovani* infection

The assessment of the intensity of infection was carried out by examining the spleen dab smears (Fig. 2.2) on day 20–25 postinfection in case of promastigotes. The hamsters were anaesthetized by injecting 890 mg/kg of thiopentone (Intraval sodium; May and Baker, United Kingdom) intraperitoneally.

The spleen biopsy was carried out aseptically through an incision of about 1 cm over the gastrosplenic region. The omentum was drawn exposing the spleen (Fig. 2.1C) and a small piece was cut and kept on a filter paper. The incision was

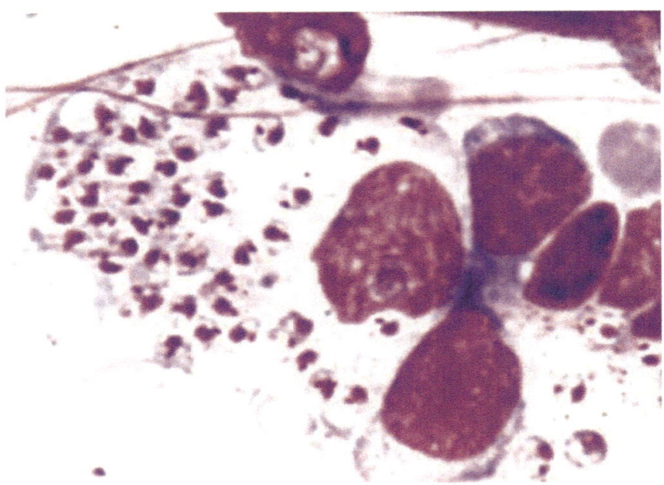

Fig. 2.2 Intracellular and nonmotile stage called amastigotes (small dotes) as seen in Giemsa-stained dab smear prepared from the spleen of an L. *donovani*-infected golden hamster.

sutured in layers. Neosporin (Burroughs Wellcome (India), Bombay) was applied on the stitched portion, and it was finally sealed with tincture benzoin. Each operated animal was treated with 100 mg/kg of neomycin sulphate orally before and after a day of biopsy. The biopsied splenic tissue was dabbed over filter paper so as to absorb excess blood and then impression smears were made on clean microscopic slides. The smears were air dried, fixed in methanol, and stained with 10% Giemsa for 45 min. The slides were examined for the number of amastigotes/100 cell nuclei [15] under the microscope. Male golden hamsters (50–709 mg) were infected intracardially with 1×10^7 amastigotes, isolated from heavily infected hamster spleens. The parasites were taken up by the macrophages of the reticuloendothelial system–the liver, spleen, and bone marrow–where they reside and multiply. A detectable level of parasite load was reached in the spleen by day 20–25 postinoculation, which was checked by conducting spleen biopsies [1]. Briefly, the hamsters were anesthetized by administering 80.0 mg/kg of thiopentine sodium intraperitoneally.

A spleen biopsy was conducted aseptically by making an incision of about 1.0 cm over the gastrosplenic region. The omentum was drawn out to expose the spleen. A small piece (approx. 2 × 2 mm) of the spleen was cut and taken out. The incision was sutured in layers. Sufficient antiseptic (neosporin antiseptic powder) was applied on the stitched portion and finally sealed with tincture benzoin. Each animal was given 100 mg/kg of neomycin sulphate orally (2 doses between 24 and 48 h) to avoid secondary infections. This treatment has no afferent effect on the leishmania parasite [16]. The biopsied spleen tissue was dabbed on a filter paper to remove excess blood and dabbed on a grease-free microscopic slide. Contact smears were air dried, fixed in absolute methanol, and stained with 10% Giemsa stain.

The stained slides were examined for the parasites and expressed as the number of amastigotes/100 cell nuclei [1,15]. Animals with 10–25 amastigotes/100 cell nuclei were included in the chemotherapeutic trials. The infected hamster with the required range of parasite burden were randomly divided into various groups of three hamsters each with assigning groups, one group for the positive control drug, second group for test compound, and third group served as untreated control. Treatment was initiated 3 days after spleen biopsies and continued for 10 consecutive days. Spleen biopsies were again conducted on day 7 posttreatment and the percent inhibition of parasites was calculated using the following formula.

$$\%\,\text{inhibition} = \text{AT} \times 100/\text{IT} \times \text{TI}$$

where AT is the actual number of amastigotes/100 spleen cell nuclei in treated animals, IT is the initial number of amastigotes/100 spleen cell nuclei in treated animals, and TI is the times increase in untreated control animals.

References

[1] S. Bhatnagar, P.Y. Guru, J.C. Katiyar, R. Srivastava, A. Mukherjee, M.S. Akhtar, M. Seth, A.P. Bhaduri, Exploration of antileishmanial activity in heterocycles : results of their in vivo and in vitro bioevaluations, Indian J. Med. Res. 89 (1989) 439.

[2] A. Dube, N. Singh, A. Saxena, V. Lakshmi, Antileishmanial potential of a marine sponge, *Haliclona exigua* (Krikpatrick) against experimental visceral leishmaniasis, Parasitol. Res. 101 (2007) 317–324.
[3] J. Sunter, K. Gull, Shape, form, function and Leishmania pathogenicity: from textbook descriptions to biological understanding [published correction appears in Open Biol. 2018 Aug;8(8):], Open Biol. 7 (9) (2017), https://doi.org/10.1098/rsob.170165, 170165.
[4] L.A. Stauber, Characterization of strains of *Leishmania donovani*, Exp. Parasitol. 18 (1) (1966) 1–11, https://doi.org/10.1016/0014-4894(66)90002-6.
[5] A. Kumar, M. Samant, P. Misra, P. Khare, S. Sundar, R. Garg, A. Dube, Immunostimulatory potential and proteome profiling of *Leishmania donovani* soluble exogenous antigens, Parasite Immunol. 37 (7) (2015) 368–375.
[6] K.P. Chang, Human cutaneous lieshmania in a mouse macrophage line: propagation and isolation of intracellular parasites, Science 209 (1980) 1240–1242.
[7] R.W. Cooper, Small species of primates in biomedical research, Lab. Anim. Care 18 (1968) 267.
[8] V.A. Dennis, R. Layan, W.L. Chapman Jr., W.L. Hanson, Leishmania donovani: Cellular and humoral immune response after primary and challenge infections in squirrel monkeys (Saimiri Sciureus), Exp. Parasitol. 61 (1986) 319.
[9] P.,.R. Anuradha, P.Y. Guru, J.C. Katiyar, N. Sethi, Successful transmission of L. donovani in Indian Iangurs (Presbytis entellus), Curr. Sci. 59 (2) (1990) 612.
[10] J.P. Haldar, S. Ghose, K.C. Saha, A.C. Ghose, Cell mediated immune response in Indian Kala-azar and post Kala-azar dermal leishmaniasis, Infect. Immun. 42 (1) (1983) 702.
[11] P.E.C. Manson-Bhar, Immunity in Kala-azar, Trans. R. Soc. Trop. Med. Hyg. 55 (1961) 550.
[12] W.E. Agu, J.P. Farrel, E.J.L. Soulsby, Complement increase in experimental Leishmania donovani infection in golden hamsters, Int. J. Parasitol. 11 (2) (1981) 113.
[13] D.S. Allain, Isolation of Trypanosoma cruzi in an acutely infected patient, J. Parasitol. 60 (1974) 526.
[14] K.P. Chang, A.L. Hendricks, Laboratory cultivation and maintenance of leishmania, in: K. P. Chang, R.S. Bray (Eds.), Leishmaniasis, Elsevier, Amsterdam, New York, Oxford, 1985, p. 214.
[15] L.A. Stauber, E.M. Franchino, J. Grun, An eight day method for screening of compounds against Leishmania donovani in the golden hamster, J. Protozool. 5 (1958) 269.
[16] S.L. Croft, Recent development in the Chemotherapy of leishmanasis, Trends Pharmacol. Sci. 9 (1988) 376.

Host immunity and immunomodulation

3

Immune response in general

There are two kinds of immunity: innate and acquired. Innate immunity is the built-in factor to resist infection. It is present before birth and not antigen-specific. The inflammatory response has no memory for previous exposure, and therefore is not enhanced by a second exposure [1]. Although needed to fight off invaders, it is not very effective without antigen-specific cells. Dr. Adriana Zeevi, a professor of pathology and surgery at the University of Pittsburgh, presented an overview of the immune system at the 2009 ASHP Midyear Clinical Meeting [2]. One of the more important components of innate immunity is phagocytosis. Phagocytes, which include neutrophils, eosinophils, macrophages, and monocytes, recognize invaders with many types of cell surface receptors that can identify pathogens. The monocytes and macrophages are less mobile but are better phagocytes. The surface receptors are a group of proteins that can recognize patterns on the pathogens such as endotoxin or lipopolysaccharide, lipoteichoic acid, and peptidoglycans. Examples of phagocytic cell proteins are C-type lectins, integrins, pentraxins, lipid transferases, and leucine-rich proteins. Phagocytes kill pathogens by engulfing the invader and then releasing stored scavengers and enzymes to finish the process. Two other components of innate immunity are the complement, which is a group of 25 proteins that can cause cell lysis when activated, and natural killer (NK) cells, which can kill viral cells or malignant cells when active immunity is not present.

Antigen-specific immunity, also known as adaptive immunity, comes from two types of cells, the B lymphocytes and the T lymphocytes. These two cells can respond to the inflammatory response that innate immunity originates and will rapidly develop a specific immune response. The B cells will then develop into a specific antibody, or immunoglobulin, that will rapidly recognize the invasion of the same antigen. This cell can in some cases confer immunity for a lifetime. This adaptive immunity can react to protein, polysaccharide, nucleic acid, and lipid invaders. Lymphocytes possess numerous cell surface proteins that have been identified by activity. The nomenclature is termed cluster of differentiation (CD). The system has identified more than 150 cell surface proteins with specific antibodies. This allows scientists to identify the cell types and their functional state of activity. A number of CD cells have been identified and used to describe a function or produce treatments. The T cells are divided into two main subgroups: CD4 (helper T cells) and CD8 (cytotoxic T cells). The CD8 cells recognize viral peptides with the major histocompatibility complex (MHC) and

Visceral Leishmaniasis. https://doi.org/10.1016/B978-0-323-91124-5.00013-8

kill the infecting cell. The CD4 cells use a different class of MHC (class II) to activate macrophages or B cells. CD3 is on all T cells but not on B cells. CD56 is located on NK cells. CD19 and CD20 are on B cells but not on T cells. Many clusters of differentiation have been identified, and new drugs will be developed to modulate these receptors. The CD20 identification on B cells was used to produce one of the more exciting cancer drugs.

Once the *Leishmania* parasite invades a host, it will try to establish infection. An inflammatory response (an innate immune response) is shown by the host due to the infection. Immune responses to protozoan infection are complex, involving both humoral and cell-mediated responses to an array of parasite antigens. Infection often involves multiple life cycle stages of these parasites, which adds to the diversity of antigens that are potentially important for the development of protective immunity. To examine the molecular basis of the immune responses elicited during these infections, recent efforts have focused on evaluating responses to defined parasite T-cell epitopes. One of the features of the genus Leishmania is the antigenic diversity that results in infection and disease progression. The severity of visceral leishmaniasis (VL) is a function of the infecting *Leishmania* species together with host genetics and consequent inflammatory and immune responses. Severity of VL has become evident from genetic and immunologic studies using the murine model [3]. The classic VL syndrome is exemplified by patients that are heavily infected throughout the mononuclear phagocyte system; develop a life-threatening disease after an incubation period of weeks to months; and have fever, severe cachexia, hepatosplenomegaly, pancytopenia (anemia, thrombocytopenia, and leucopenia, with neutropenia, marked eosinopenia, and relative lymphocytosis and monocytosis), and hypergammaglobulinemia (mainly IgG from polyclonal B-cell activation) with hypoalbuminemia. These situations often occur with poor people. Hence, VL is linked to poverty, economic development, and various environmental changes such as deforestation, urbanization, migration of people into endemic areas, building of dams, etc. Therefore, a great deal of attention has been paid to VL during the past decade because of etiological and behavioral factors, the resistance of the parasite to drugs, and the socioeconomic conditions of the population at risk because untreated VL disease is always fatal. VL is a life-threatening form of leishmaniasis involving the whole reticuloendothelial system, including organs such as bone marrow and skin where drug delivery becomes difficult. In this disease, the macrophages, which are normally destined to kill any invading or foreign organism, become the site not only for lodging the parasite but also for its multiplication. This results in the depression of the immune status of the host, which drastically alters the immune system leading to derangement of various physiological functions. These conditions favor a suitable situation for intercurrent infections to further worsen the situation, and the management of such a condition becomes more complex with existing drug therapy. The increasing failure of previously effective drugs and inherent serious toxicities associated with the drug. Therefore, the development of efficacious, safe, inexpensive, and short-course treatment is urgently required for better management of Kala-azar.

Host-specific immune response in VL

Macrophage activation by T-cell-derived cytokines is required to establish control of a cellular infection and a progressive disease. In these cells, *Leishmania* growth inhibition or killing is mediated by the production of nitric oxide from inducible nitric oxide synthase (iNOS). The transcription of this enzyme is induced and enhanced by interferon IFN-γ, a key cytokine in *Leishmania* control, and by other synergic cytokines such as tumor necrosis factor (TNF-α) or IL-2 [4]. Macrophage activation results from a network of up regulating or downregulating chemokine signals, which modulate immune cell proliferation, activation, and expansion. VL affects humans; its popular experimental model is a mouse or a hamster. Therefore, the immune response to VL is discussed below in the human, mouse, and hamster.

Human

VL is a potentially fatal disease endemic in parts of Sudan, India, Nepal, Bangladesh, and Brazil. The majority of people in endemic regions are asymptomatic to *L. donovani* infection and develop life-long immunity against reinfection. Human VL caused by *L. donovani* or *L. infantum* is a severe disease with generalized spread of the parasites to the reticuloendothelial system, such as the spleen, liver, and bone marrow. In communities exposed to infection, individuals develop a strong cellular immunity in an age-related fashion, as shown by cross-sectional studies on reactivity to intradermal leishmanial antigens (leishmanin skin test, LST) [5]. Subclinical or asymptomatic cases identified by low-titer anti-*Leishmania* antibodies may show anergy to intradermal antigens and develop the disease thereafter, or convert to seronegative and become LST-positive, suggesting spontaneous healing [6]. During acute VL, patients are unresponsive to LST, whereas 6–12 months from successful treatment, the test converts to positive. Both spontaneous and drug-induced healing are thought to be followed by strong protective immunity. There are several studies indicating that a Th2-type response predominates during the acute disease, such as the suppression of T-cell reactivity to *Leishmania* antigens, the predominance of endogenous IL-4 over IFN-γ production, and polyclonal B-cell activation resulting in hypergammaglobulinemia [7]. In contrast with the *L. major* mouse model, however, other studies indicate that both Th1- and Th2-like cells appear to be activated during the course of the disease, as revealed by the simultaneously high levels of IFN-γ and IL-10 detected in patients [8]. Furthermore, the upregulation of IL-10, rather than IL-4, appears to be a constant feature in the clinical disease. This cytokine has been associated with immune suppression commonly seen in VL patients [9] and to the development of Post Kala-azar Dermal Leishmaniasis (PKDL: a skin manifestation caused by *L. donovani* after apparent VL healing [10]. Taken together, the above findings provide evidence for the existence of a Th1/Th2 dichotomy in the T-cell response in human leishmaniases. Interestingly, the outcome of the infection appears to be determined and regulated by the balance between the two parasite-specific T-cell populations [11].

In humans, the measurements of cytokines in culture supernatants of *Leishmania* antigen-activated peripheral blood mononuclear cells (PBMC) and T-cell clones have helped in determining the stimulation of Th1 and Th2 immune responses. Studies of tissue cytokine mRNA expression reveal a role for IL-10 in downregulating $CD4^{+}T$ cell responses and the involvement of IL-10 in the disease pathology of *L. donovani* infections. However, active VL also finds correlation with enhanced induction of IFN-γ, IL-2, IL-10, and IL-4. After cure levels of IFN-γ, IL-4 and IL-10 persist, suggesting a coexistence of Thl and Th2 in kala azar patients as well as in cured individuals [12]. The dichotomous Thl/Th2 cellular response seen in human VL is probably regulated by the balance of the parasite-specific cellular populations [11], which may play an important regulatory role in determining the outcome of the infection in humans. Thus, even in humans it is difficult to demarcate the responses leading either to visceral disease or to protective immunity with *L. donovani*.

Individuals with overt VL display a negative skin test response to *Leishmania* antigens. Furthermore, PBMCs from such individuals fail to proliferate or to produce IFN-γ when exposed to a specific antigen in vitro [13]. Interestingly, the failure of PBMCs from recently infected infants to produce this cytokine is predictive of the evolution of the disease toward classical patent VL, whereas conversely, the production of IFN-γ in vitro is indicative of the maintenance of the infection at a subclinical level [14]. These observations point to the critical importance of early T cell activation patterns in determining the eventual outcome of the infection.

Which mechanisms underlie the poor T-cell reactivity to *Leishmania* antigens characteristic of patients with active disease? Much of this effect might be due to the unbalanced production of IL-10. Indeed, elevated blood levels of IL-10 as well as high production of this cytokine (or its mRNA) by lymph node cells or PBMC from kala azar patients appear to be a hallmark of this form of infection [8,9,15], whereas cell populations collected from cured individuals fail to express this immunomodulatory molecule. IL-10 therefore appears to constitute a major regulatory cytokine whose production may critically control the outcome of infection. Failure to produce IL-12 has similarly been associated with the active form of the disease [9]. It is also suggested that unbalanced IL-10 production might play a role in the progression of the disease toward the cutaneous form of the infection, known as PKDL [9], which is observed in a high proportion of patients. Interestingly, PKDL appears to develop as leishmania-reactive T cells start to become detectable in peripheral blood following the cure of the visceral infection. It is also observed in patients whose cells respond to the *Leishmania* antigen in vitro by high IFN-γ release. Furthermore, a histological examination of the lesions suggests that the development of skin lesions characteristic of PKDL involves an inflammatory component [16].

Mouse

Different *Leishmania* species cause clinically distinct diseases and the severity of the disease caused by any given parasite can vary markedly between individual hosts [17]. So far, two host systems have been classified for studying *Leishmania* infection on the basis of the susceptibility and resistance of the host. Murine models for experimental leishmaniasis are well established. Parasites are injected underneath the skin of the footpad. Most of the mice strains such as C57BL/6, CBA/J, C3H, or BIOD2 resist the infection, with a clinical cure within a few months ([18]). BALB/c and all T-cell immunodeficient strains manifest systemic VL, leading to death [19]. Resistance and susceptibility are closely related to the development of the T-cell responses of the Th1 or Th2 type, respectively.

It is well documented that the Th1 immune response is the key event to prevent *Leishmania* infection. Activated Th1 cells induce IFN-γ that in turns activates the macrophages, killing the parasites. C57BL/6 mice mount an early Th1 immune response and prevent the further growth of the parasite, causing a self-healing phenotype. A susceptible BALB/c strain mounts an early Th2 response, resulting in nonhealing lesions and the exaggeration of the disease [20]. The respective resistance and susceptibility of C57BL/6 and BALB/c strains depend not only on the Th1 and Th2 types of immune response of CD4+ T cells, but also on the genetic background of the host.

In general, the immune responses following the infection of inbred mouse strains with viscerotropic *Leishmania* species, such as *L. donovani* or *L. infantum*, are similar to those observed in the *L. major* mouse model. However, BALB/c mice do not appear to exhibit a similarly high susceptibility to these parasites, as the intravenous injection of visceral *Leishmania* results in a self-healing chronic infection. Furthermore, cytokine phenotypes elicited by viscerotropic *Leishmania* in this mouse model are not typical of a Th2-type response [21].

One of the most important advances in the last decade is our understanding of the Thl/Th2 paradigm, which has been best studied in the mouse model of *L. major*. The differential development of Th1 and Th2 cells in mice infected with *L. major* has been used to define the factors that regulate the development of $CD4^{+}$T-cell subsets. While control of the infection in resistant mice such as C3H and C57BL/6 is associated with the expansion of a Th1 subset of the CD4 lymphocyte that secretes macrophage-activating cytokines such as IFN-γ and IL-2 [22], susceptibility is instead linked to the expansion of the CD4 lymphocytes belonging to the Th2 subset that secrete IL-4, IL-5, IL-l, and IL-13 [20]. However, such a polarized Th1 or Th2 response has not been observed in murine models of *L. donovani* infections. Studies of C57BL/10 and B10-D2/N mice infected with *L. donovani*, the curing and noncuring models of VL, respectively, have shown that the Th2 type cytokine, IL-4, is not required for the susceptible phenotype to manifest [21]. However, cytokines associated with Thl and Th2 cells have been reported in BALB/c mice infected with *L. donovani* [23]. BALB/c mice infected with parasites of *L. donovani* develop either an initial susceptibility with spontaneous acquisition of resistance and control over parasite replication [23], or a progressive visceral disease mimicking infections of *L. donovani* in humans [24]. While the control of infection in the former model is known to be partly dependent on Thl cell responses [25,26], the characteristics of the T cell responses in the latter have not been identified. This is a challenge for the development of a vaccine. To achieve a detectable visceral disease, often mice have to be challenged with a very large number of parasites, which is far from those inoculated during a natural infection. Recently, an intradermal murine model of VL was explored for the establishment of a chronic infection pattern in susceptible BALB/c mice resembling a course of disease similar to that of human VL [27]. This model could serve as an important tool in future vaccine studies against VL.

Hamster

The golden hamster is an experimental model of progressive visceral disease characterized by symptoms such as hyperglobulinemia, anemia, and cachexia, leading to the deaths of the animals. Interestingly, an inoculum (10^3–10^5) of *L. infantum* promastigotes could result in the development of both symptomatic and asymptomatic states with a spontaneous protective response [28]. The fact that even a small inoculum that is close to the numbers seen during natural infection could lead to disease [29] along with the unpredictable nature of the immune state suggest that the various clinical and pathogenic features exhibited in hamsters are very similar to those in human VL, making it an excellent model of VL.

Recent studies have shown that the mice model of *L. donovani* does not reproduce the features of active human VL such as chronic fever, hepatosplenomegaly, pancytopenia, and profound cachexia; it also has an ineffective antileishmanial cellular response [30]. On the contrary, the Syrian golden hamster model of active VL is closely related to the human counterpart, as shown by the relentless increase in the visceral parasite burden, progressive cachexia, hepatosplenomegaly, pancytopenia, hypergammaglobulinemia, and ultimately death. While the mice are either intrinsically resistant or susceptible to *Leishmania* infection and offer a well-characterized genetic makeup, chiefly by the use of inbred, recombinant, and naturally or experimentally mutated strains, hamsters provide an excellent model for an overtly susceptible host. Therefore, hamsters are used for histopathological studies, drug efficacy studies, and vaccine studies despite the lack of fine immunochemicals, which limits the mechanistic exploration of immune responses to *Leishmania* infection [31].

Immunomodulation and immunomodulators

In early 1940, sodium antimony gluconate was developed to treat VL, but resistance was reported after a few years of use. After that, a few antileishmanial drugs were also developed, but they are not satisfactory due to cytotoxicity and other issues [32]. The new therapeutic intervention should be a drug-based approach to manipulate the immune system. Therefore, there is a need for the development of immunomodulators that do immunomodulation in the host. Immunomodulation denotes a process in which the host immune response is altered to a desired level. Parasites are also capable of modulating the response of the host immune system to their presence to establish an infection. Immunomodulation in the host is done due to disease progression or therapeutic intervention. But the discovery proceeded slowly toward the production of immunomodulators. The goal is to move toward the development of tolerance while still defeating any infection that might arise. The problem with this type of therapy is that the immune system is redundant and cannot be defeated as easily as one would think. The biopharmaceutical science of immunomodulating agents was discussed at the ASHP meeting by Dr. Raman Venkatarammanan, also a professor at the University of Pittsburgh [33]. Pharmacologic mechanisms to alter immune function have come a long way. Many of the drugs that mediate immune function are not used for this purpose (captopril, heparin, and nonsteroidal antiinflammatory drugs (NSAIDs). Many of the early drugs were of the type that moderate immunity by blocking the immune components of innate immunity. These agents have been shown to have effective immunomodulation.

Therapeutic agents that do immunomodulation are known as immunomodulators. They are a diverse array of recombinant, synthetic, and natural preparations, often cytokines. They can be in the form of small molecules, large molecules, peptides, and many different types of proteins. Elements that must be taken into consideration are the same as in any drug that goes into the human body. Metabolic, absorptive, distribution, and excretion barriers must be taken into account when using these medications. The bioavailability of the drugs approved by the US Food and Drug Administration can run from 15% for sirolimus to as much as 95% for mycophenolate mofetil. The absorption peak can be as low as 1 h for mycophenolate and as high as 6 h for leflunomide. This information is useful when switching doses from the intravenous (IV) route to the oral route. The medications (leflunomide and mycophenolate) that have a high bioavailability can be dosed similarly when switching from IV to oral routes and produce a somewhat low variability in effect. By contrast, sirolimus and tacrolimus, which have low bioavailability, will require higher oral doses when a switch is made from parenteral to oral dosage. Most of the above-mentioned drugs can have some effect in all disease states. By using those drugs, we can think with immunomodulators for VL treatment. The therapeutic use of this intervention is called "immunomodulation," when we use medications affecting the immune system. Immunomodulators may be of the following types:

1. Immunostimulatory/immunopotentiator: immunostimulating/restoring effects of antibacterial agents are beneficial in the common disease.
2. Immunosuppressive/immunological tolerance: immunosuppressive antimicrobials have shown promise in inflammatory diseases.

The definition of immunomodulation refers to the action undertaken by the medication on autoregulating processes that steer the immunological defense system. A lot of antihomotoxic medications intervene, and they are more than useful with proven therapeutic action. Most of the clinically used immunostimulants/immunosuppressants are cytotoxic drugs, which have serious side effects. Therefore, researchers have growing interest in using herbal medicines as multicomponent agents to modulate the complex immune system in the prevention of infections. Many therapeutic effects of plant extracts have been suggested to be due to their wide array of immunomodulatory effects and influence on the immune system of the human body. Phytochemicals such as alkaloids, flavonoids, polysaccharides, lactones, diterpenoids glycosides, etc., are present in several plants. They have been reported to be responsible for the plant immunomodulating properties. Thus, the search for natural products of plant origin as new leads for potent and safe immunomodulators is gaining much major research interest in the direction of a new molecule search for VL therapy. The next chapter explains the immunomodulatory activity of natural products.

References

[1] D. Zheng, T. Liwinski, E. Elinav, Interaction between microbiota and immunity in health and disease, Cell Res. 30 (6) (2020 Jun) 492–506.

[2] Biopharmaceutical aspects of immunomodulataroy agents, Presented at the ASHP Annual Meeting in Atlanta, December 2002, https://www.myast.org/sites/default/files/SECRETARY.pdf#.

[3] B.M. Scorza, E.M. Carvalho, M.E. Wilson, Cutaneous Manifestations of Human and Murine Leishmaniasis, Int. J. Mol. Sci. 18 (6) (2017) 1296. Published 2017 Jun 18 https://doi.org/10.3390/ijms18061296.

[4] M. Belosevic, D.S. Finbloom, M.S. Meltzer, C.A. Nacy, IL-2. A cofactor for induction of activated macrophage resistance to infection, J. Immunol. 145 (1990) 831–839.

[5] S. Bettini, M. Gramiccia, L. Gradoni, E. Pozio, S. Mugnai, M. Maroli, Leishmaniasis in Tuscany (Italy): VIII. Human population response to leishmanin in the focus of Monte Argentario (Grosseto) and epidemiological evaluation, Ann. Parasitol. Hum. Comp. 58 (1983) 539–547.

[6] C.R. Davies, A.S. Mazloumi Gavgani, Age, acquired immunity and the risk of visceral leishmaniasis: a prospective study in Iran, Parasitology 119 (1999) 247–257.

[7] A. Kharazmi, K. Kemp, A. Ismail, et al., T-cell response in human leishmaniasis, Immunol. Lett. 65 (1999) 105–108.

[8] E. Cillari, G. Vitale, F. Arcoleo, P. D'Agostino, C. Mocciaro, G. Gambino, R. Malta, G. Stassi, C. Giordano, S. Milano, S. Mansueto, In vivo and in vitro cytokine profiles and mononuclear cell subsets in Sicilian patients with active visceral leishmaniasis, Cytokine 7 (1995) 740–745.

[9] H.W. Ghalib, M.R. Piuvezam, Y.A. Skeiky, M. Siddig, F.A. Hashim, A.M. El-Hassan, D.M. Russo, S.G. Reed, Interleukin 10 production correlates with pathology in human *Leishmania donovani* infections, J. Clin. Invest. 92 (1993) 324–329.

[10] S. Gasim, A.M. Elhassan, A. Kharazmi, E.A. Khalil, A. Ismail, T.G. Theander, The development of post-kala-azar dermal leishmaniasis (PKDL) is associated with acquisition of Leishmania reactivity by peripheral blood mononuclear cells (PBMC), Clin. Exp. Immunol. 119 (3) (2000) 523–529.

[11] M. Kemp, T.G. Theander, A. Kharazmi, The contrasting roles of CD4+ T-cells in intracellular infections in human: Leishmania as an example, Immunol. Today 17 (1996) 13–16.

[12] S. Sunder, S.G. Reed, S. Sharma, A. Mehrotra, H.W. Murray, Circulating T helper 1 (Th1) cell and Th2 cell associated cytokines in Indian patients with visceral leishmaniasis, Am. J. Trop. Med. Hyg. 56 (5) (1997) 522.

[13] E.M. Carvalho, W.D. Johnson, E. Barreto, P.D. Marsden, J.L. Costa, S. Reed, H. Rocha, Cell mediated immunity in American cutaneous and mucosal leishmaniasis, J. Immunol. 135 (1985) 4144–4148.

[14] E.M. Carvalho, A. Barral, D. Pedral-Sampaio, M. Barral-Netto, R. Badaro, H. Rocha, Immunologic markers of clinical evolution in children recently infected with *Leishmania donovani* chagasi, J Infect Dis 165 (1992) 535–540.

[15] B.J. Holaday, M.M. Pompeu, T. Evans, D. DeBraga, M.J. Texeira, S.A. de Queiroz, M.D. Sadick, A.W. Vasconcelos, J.S. Abrams, R.D. Pearson, R.M. Locksley, Correlates of *Leishmania*-specific immunity in the clinical spectrum of infection with *Leishmania chagasi*, J Infect Dis 167 (1993) 411–417.

[16] A. Ismail, A.M. El Hassan, K. Kemp, S. Gasim, A.E. Kadaru, T. Moller, A. Kharazmi, T.G. Theander, Immunopathology of post kala-azar dermal leishmaniasis (PKDL): T-cell phenotypes and cytokine profile, J. Pathol. 189 (4) (1999) 615–622.

[17] D.J. Bradley, Genetic control of natural resistance to *Leishmania donovani*, Nature 250 (1974) 353–354.
[18] E. Handman, *Leishmania* vaccines: Old and new, Parasitol. Today 13 (6) (1997) 236.
[19] J.G. Howard, C.E. Hale, W.L. Chan-Liew, Immunological regulation of experimental cutaneous leishmaniasis. I. Immnogenetics aspects of susceptibility to *Leishmania tropica* in mice, Parasite Immunol. 2 (1980) 303–314.
[20] H. Himmelrich, P. Launois, I. Maillard, T. Biedermann, F. Tacchini-Cottier, R.M. Locksley, et al., In BALB/c mice, IL-4 production during the initial phase of infection with *Leishmania major* is necessary and sufficient to instruct Th2 cell development resulting in progressive disease, J. Immunol. 164 (2000) 4819–4825.
[21] P.M. Kaye, A.J. Curry, J.M. Blackwell, Differential production of Th1- and Th2-derived cytokines does not determine the genetically controlled or vaccine-induced rate of cure in murine visceral leishmaniasis, J. Immunol. 146 (1991) 2763–2770.
[22] S.L. Reiner, R.M. Locksley, The regulation of immunity to Leishmania major, Annu. Rev. Immunol. 13 (1995) 151–177.
[23] G.D. Miralles, M.Y. Stoeckle, D.F. McDermott, F.D. Finkelman, H.W. Murray, Induction of Th1 and Th2 cell-associated cytokines in experimental visceral leishmaniasis, Infect. Immun. 62 (1994) 1058–1063.
[24] F. Afrin, N. Ali, Adjuvanticity and protective immunity elicited by *Leishmania donovani* antigens incapsulated in positively charge liposomes, Infect. Immun. 65 (6) (1997) 2371.
[25] H.W. Murray, J. Hariprashad, R.L. Coffman, Behavior of visceral *Leishmania donovani* in an experimentally-induced Th2 cell-associated response model, J. Exp. Med. 185 (1997) 867–874.
[26] H.W. Murray, K.E. Squires, G.D. Miralles, M.Y. Stoeckle, A. Granger, A. Piperno, C. Bogdan, Acquired resistance and granuloma formation in experimental visceral leishmaniasis: differential T cell and lymphokine roles in initial vs. established immunity, J. Immunol. 148 (1992) 1858–1865.
[27] S. Ahmed, M. Colmenares, L. Soong, K. Goldsmith-Pestana, L. Munstermann, R. Molina, D. McMahon-Pratt, Intradermal infection model for pathogenesis and vaccine studies of murine visceral leishmaniasis, Infect. Immun. 71 (1) (2003) 401–410.
[28] J.M. Requena, C. Alonso, M. Soto, Evolutionarily conserved proteins as prominent immunogens during *Leishmania* infections, Parasitol. Today 16 (2000) 246–250.
[29] A. Warburg, Y. Schlein, The effect of post-bloodmeal nutrition of *Phlebotomus papatasi* on the transmission of *Leishmania major*, Am. J. Trop. Med. Hyg. 35 (5) (1986) 926–930.
[30] P.C. Melby, G.B. Ogden, H.A. Flores, W. Zhao, C. Geldmacher, N.M. Biediger, S.K. Ahuja, J. Uranga, M. Melendez, Identification of vaccine candidates for experimental visceral leishmaniasis by immunization with sequential fractions of a cDNA expression library, Infect. Immun. 68 (10) (2000) 5595.
[31] P.C. Melby, J. Yang, W. Zhao, L.E. Perez, J. Cheng, *Leishmania donovani* p36 (LACK) DNA is highly immunogenic but not protective against experimental visceral leishmaniasis, Infect. Immun. 69 (8) (2001) 4719–4725.
[32] A. Kumar, S.C. Pandey, M. Samant, Slow pace of antileishmanial drug development, Parasitol. Open 4 (2018) 1–11. e4.
[33] R. Venkataramanan, Biopharmaceutical considerations in using new immunomodulating agents, Presented at the 37th American Society of Health-System Pharmacists Midyear Clinical Meeting and Exhibits; Atlanta, Georgia, 2002.

VL and natural products

4

VL: A general account

Visceral leishmaniasis (VL) is largely prevalent in developing countries, where the rural population is the most vulnerable target. It has had a long history in India since its original identification in the middle of the 19th century as Dumdum or Burdwan fever, which is distinct from malaria [1]. Since then, the disease has been endemic in eastern regions of the Indian subcontinent, including Assam (the Bramhaputra basin), the Bay of Bengal (West Bengal and Bangladesh), and Bihar (the Gangetic basin). After World War II, the large-scale spraying of DDT in connection with the malaria control program in India and in Bangladesh (then East Pakistan) resulted in the virtual disappearance of kala azar as a public health problem on the subcontinent [2]. After a dormant period of more than 20 years, there was a dramatic resurgence of the disease in Bihar with two major epidemiological peaks, one in 1977–78 and the other in the early 1990s. Since then, Bihar continues to be a highly endemic area for kala azar [3]. The disease has also been reported from different areas of the subcontinent, mostly those adjoining Bihar, and its spread is probably due to population migration. The situation was further aggravated due to the occurrence of cases unresponsive to antimonials [4]. About 30%–50% patients do not respond to the prevailing regimen of antimonials. Resistance to antimonials is a critical issue in the prolonged kala azar epidemic in India, where 37%–70% of the estimated 250,000 annual cases are currently reported to be nonresponsive to this first-line agent. In fact, clinical resistance appears to be local to Bihar, where 65% of the patients fail to respond to treatment compared with 11% in the neighboring state of Uttar Pradesh, despite the use of the same drug lot at both treatment sites [4]. This has led groups to ask whether Sb^V should continue to be used as well as inquire about the development of new modes of therapy. Although the exact cause of the resistance remains elusive, there is evidence that it is intrinsic to the parasite and not to the host. By means of a model of intracellular infection, a strong correlation between clinical response and *L. donovani* antimony sensitivity was found in vitro [5].

Immune suppression has been found to be involved in the etiopathogenesis of VL as the metabolic system gets disturbed, and ultimately the deterioration of human functional capacity takes place [6]. Such immunosuppression has been found in infections caused by protozoan parasites [7]. To prevent infection, procedures of revitalization and rejuvenation could be adopted to increase the power of resistance (increased immunity) in humans against the disease. Thus, vitalizers that appear akin to an antiparasitic agent of plant origin (whether from Ayurveda, Chinese, or Russian traditional medicine) may act as inducers of interferon for fighting infection. Chemotherapy has huge side effects and is costly.

The immune system, which defends against infections, includes the cellular response (for instance, leukocytes or white blood cells) and the humoral

Visceral Leishmaniasis. https://doi.org/10.1016/B978-0-323-91124-5.00014-X

response (antibodies). For immunodeficiency or increased immunity, immunomodulators are used to provide a normal immune response. Immunomodulators, also known as biological response modifiers (BRMs) are substances that modify the immune response. There is evidence that biological products from food and other sources enhance the action of BRMs produced by the immune system. The immunomodulatory agents act as general tonics to increase energy and prevent disease, which is related to its effect on the immune system [8]. The leaves, bark, and fruit of plants are used in the indigenous system of medicine for an antimicrobial effect. Phytochemical studies revealed many chemical groups such as terpenes, flavonoids, phenolic materials, etc. [9]. The pharmacological activity was evaluated and it showed antioxidant activity, central nervous system depressant activity, and antiinflammatory activity.

Natural products as anti-VL agent

Efforts to find new chemotherapeutics for leishmaniasis have been ongoing for decades, and numerous potential drug candidates and/or putative drug targets have been proposed. No new drugs have become available to VL patients except the oral drug miltefosine, which was recently approved in India for VL treatment [10]. Therefore, the continuation of drug-screening efforts must remain a priority, partly because resistance to first-line drugs is growing. The lack of an effective antileishmanial drug has caused a renewed interest in the study of medicinal plants as sources of new chemotherapeutic compounds with better activities and fewer side effects. Many people who live in areas where leishmaniasis is endemic rely on traditional medical systems for treatment. In most cases, the therapy consists of the oral administration of plant extracts for the systemic forms of the disease and of topical preparations for the cutaneous forms of infection. Due to the limited availability and affordability of pharmaceuticals, the majority of world's population depends upon traditional medical remedies from plant sources. It is estimated that more than 20,000 species of higher plants are used for their medicinal value throughout the world. Many well-known drugs such as quinine (extracted from the bark of the *Cinchona* tree) that have natural origin are used to treat malaria. The natural molecule quinine was followed by the subsequent development of the synthetic derivatives chloroquine, amodiaquine, primaquine, and mefloquine. More recently, artemisinin gained wider recognition as an antimalarial drug that is isolated from the herb *Artemisia annua*. Established human antiprotozoal drugs from natural resources include emetine from *Psychotria ipecacuanha* (used in treating amoebiasis). Therefore, there should be increased awareness of the potential of natural products, which may lead to the development of much-needed new antileishmanial drugs. There are many problems with the currently available drugs for treating VL. Many of these drugs are poorly tolerated because of their side effects. Other problems include limited availability, prohibitive cost, increasing drug resistance, etc. [11]. Therefore, a new, cheap, safe, and effective antiparasitic drug from natural origins is urgently needed for tropical diseases such as VL. More than 1300 new drugs were introduced in the last 25 years of the 20th century, but unfortunately only 13 were for tropical diseases [11]. Some natural products have shown some antileishmanial potential, including:

Diospyrin: It is a derivative of bis-naphthoquinone isolated from the bark of *Diospyros montana*. It is reported to be active against *L. donovani* promastigotes with a MIC of 1 μg/mL [12].

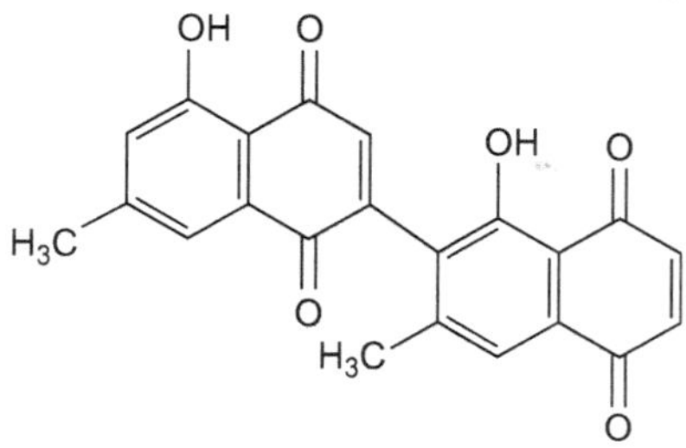

PX-6518: It is a glycosylated saponin whose antileishmanial activity was detected in a screen of some 10,000 plant extracts. The chemical structure and antileishmanial activity of the six triterpene saponins in the PX-6518 mixture were isolated from the leaves of *M. balansae* [13]. This case illustrates the difficulties in pursuing natural products, which are often chemically complex and provide few opportunities for rapidly investigating SAR by synthesizing or procuring analogues.

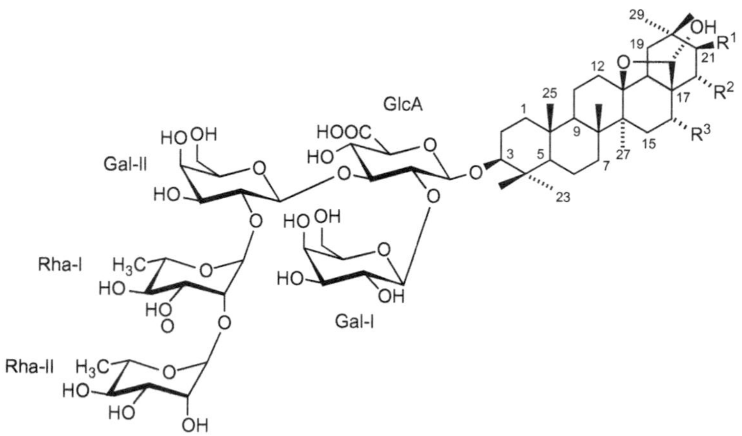

Plumbagin: It is a naphthoquinone isolated from the genus *Plumbago*. It has the anti-amastigote activity of *L. donovani* and the reported IC50 value was 0.42 μg/mL [14].

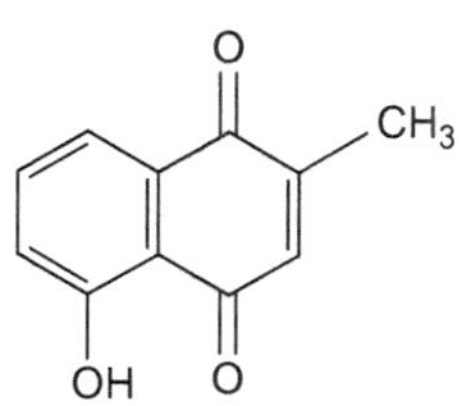

Lapachol: It is a prenylated hydroxynaphthoquinone isolated from a species of *Tecoma* (*Bignoniaceae*). It was active against amastigotes of *L. donovani,* but this activity was not very strong during the examination of peritoneal mice macrophages infected with *L. donovani* [15].

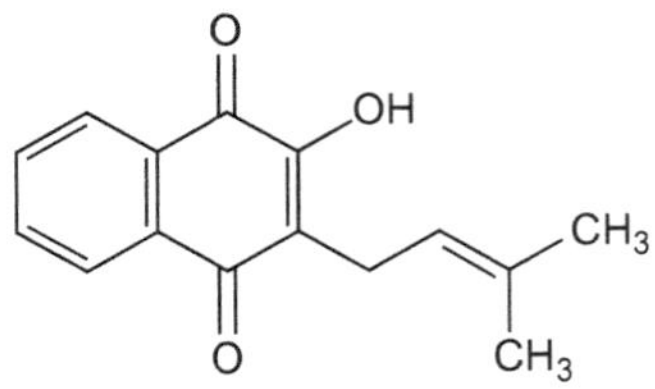

4-hydroxy-1-tetralone: It is a metabolite isolated from the bark of the Ulmaceae family plant *Ampelocera edentula.* It is known as an intermediate in the biosynthesis of 1,4-naphthoquinones. It was reported as an active natural product against the promastigotes of *L. donovani* with an IC90 of 10 μg mL^{-1} [16].

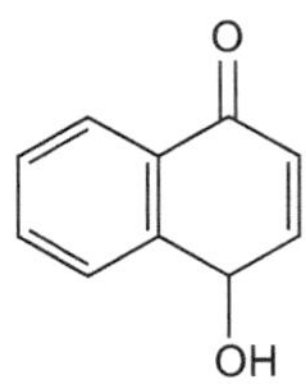

Dihydroquinone hydropiperone: Its prenylated compound was isolated from *Peperomia galioides* (Piperaceae). It has shown deleterious effects against *L. donovani* promastigote, causing a total lysis of the parasites at a concentration of 100 μg mL^{-1} [17].

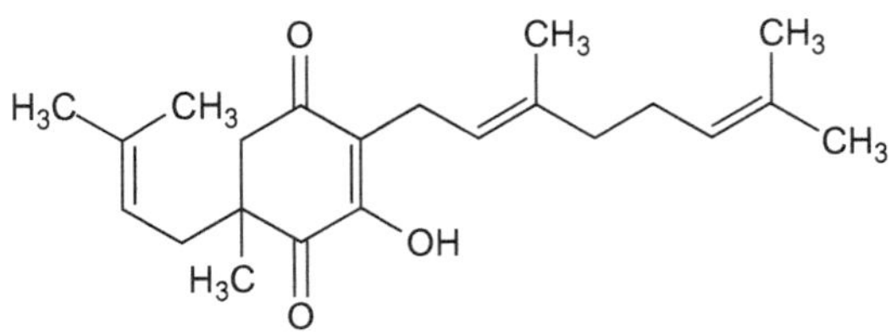

Aloe-emodin: It is an anthraquinone isolated from the aerial parts of *Stephania dinklagei* (Menispermaceae). Good leishmanicidal activity was reported against promastigotes and amastigotes of *L. donovani* at **IC50** values of 185.1 and 90 μM, respectively [18].

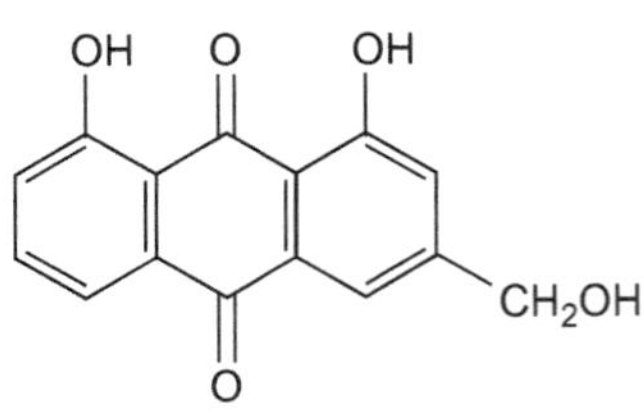

Isoguattouregidine: It is an isoquinolinic alkaloid, isolated from the bark of *Guatteria foliosa* (Annonaceae). It caused the total lysis of *L. donovani* parasites when evaluated at a concentration of 100 μg mL^{-1} [19].

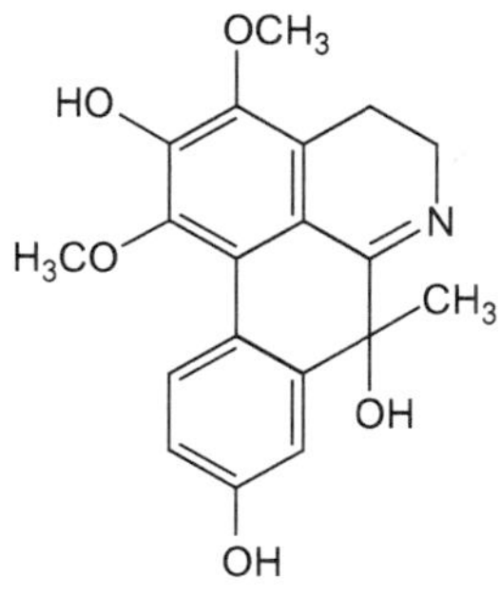

Anonaine: Another alkaloid from the trunk bark and roots of *Annona spinescens* (Annonaceae) was evaluated; it caused total lysis of the *L. donovani* parasites [20].

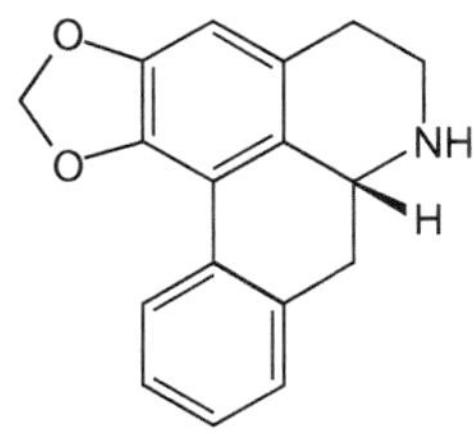

Liriodenine: Annona spinescens (Annonaceae) is the source of the alkaloid Liriodenine. It has been reported to show antileishmanial activity against the promastigotes of *L. donovani* [20]. The same compound from the stem bark of *Rollinia emarginata* (Annonaceae) exhibited leishmanicidal activity with an IC100 of 100 μg mL^{-1} [21].

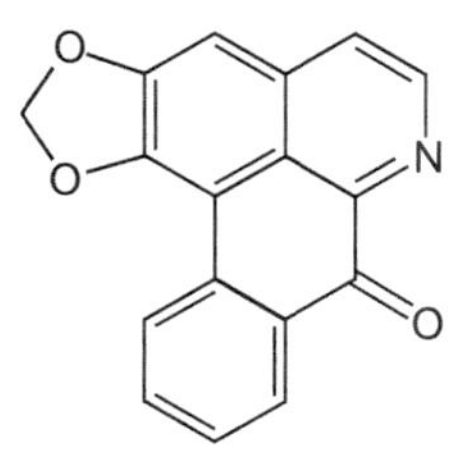

Gyrocarpine: It has shown strong leishmanicidal activity in vitro (10 μg mL^{-1}) against the promastigote forms of *L. donovani* [22].

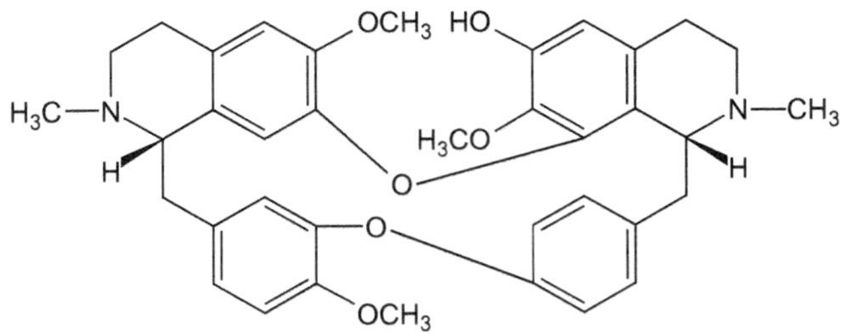

Tetradrin: It is a bisbenzylisoquinolinic alkaloid that was reported with leishmanicidal activity at 10 μg mL^{-1} against the *L. donovani* promastigote [22].

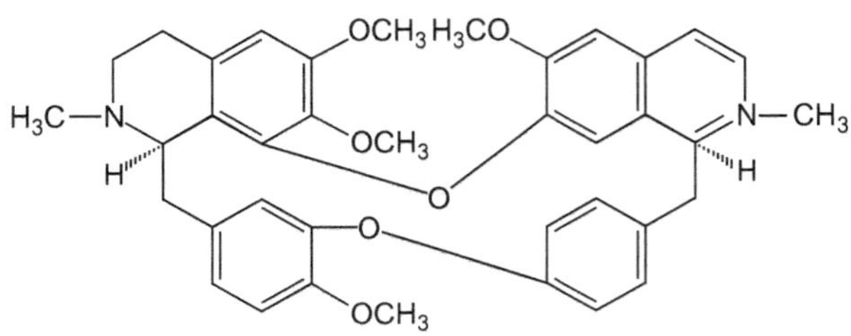

Chimanine D: It has strong in vivo action against *L. donovani* parasites [16].

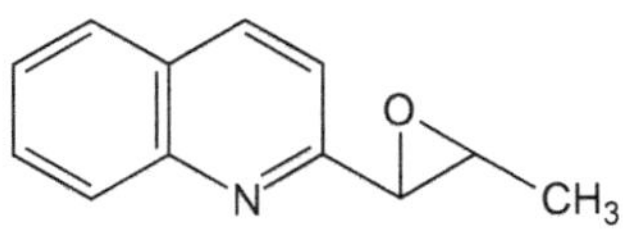

Indole alkaloids: Harmane, pleiocarpine, and buchtienine, isolated from the stem bark and leaves of *Kopsia griffithii* (Apocynaceae), have been reported to possess toxic activity against promastigotes of *L. donovani* [23].

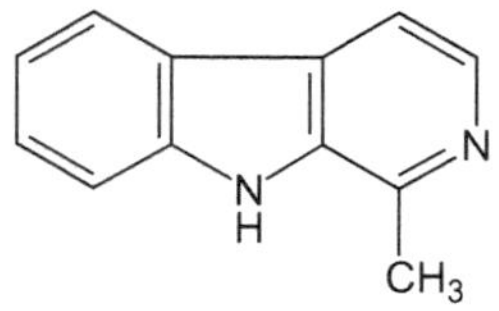

Harmoane

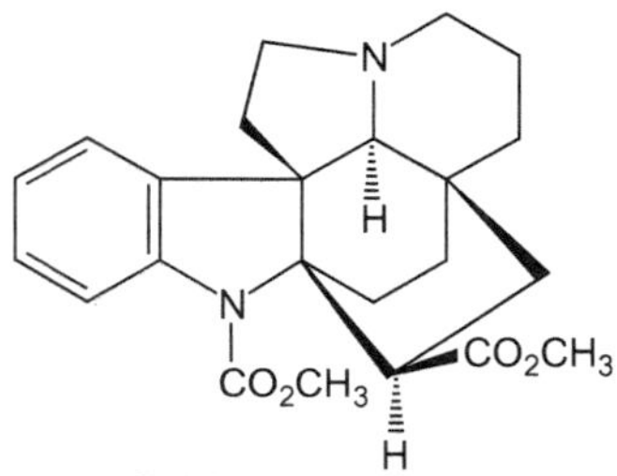

Ppleiocarpine

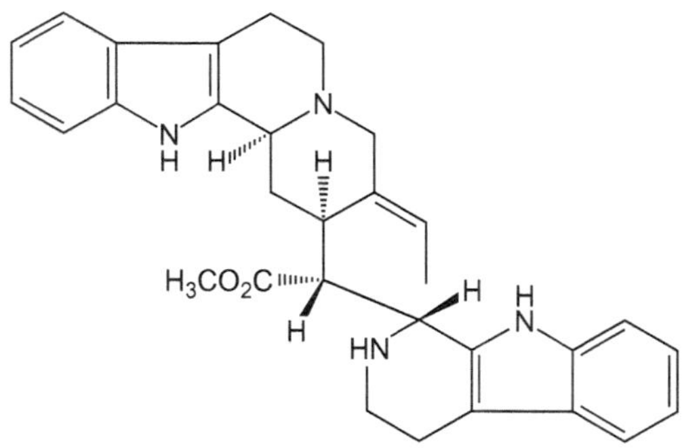

Buchtienine

Sarachine: It is a steroidal alkaloid isolated from the *Saracha punctata* (Solanaceae) Bolivian plant leaf that completely inhibits the growth of the *L. donovani* promastigote at a concentration of 10 μg mL^{-1}.

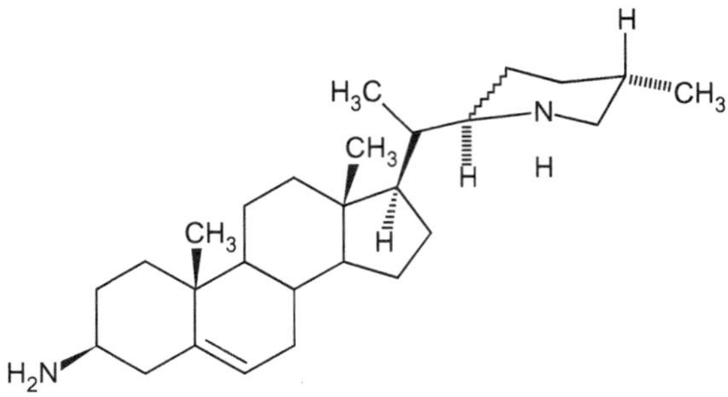

Holamine (R = H) and holacurtine (R = OH): They were isolated from *Holarrhena curtisii* (Apocynaceae) and exhibited leishmanicidal activity against promastigotes of *L. donovani* [24].

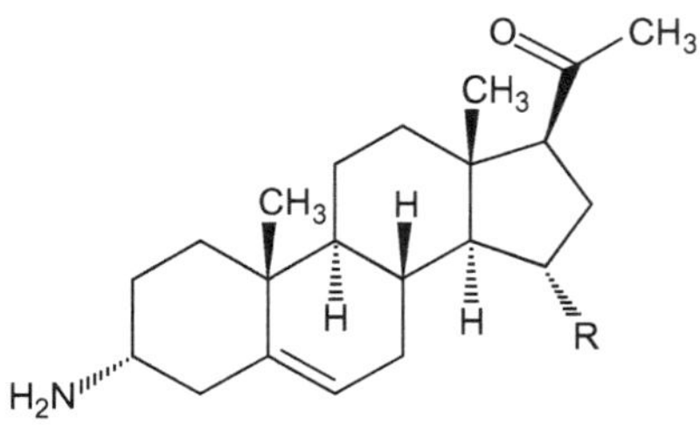

Other alkaloids: Daphnandrine, obaberine, gyrocarpine, limacine, piperine, benzoxazol-2(3*H*)-one, etc., have also shown good leishmanicidal activity.

Arbortristosides (R^1 =p-methoxycinnamoyl) and 6-β-hydroxyloganin (R^1 =H): They were isolated from the seeds of *Nyctanthes arbortristis* (Oleaceae) and possess in vitro activity against the *L. donovani* amastigote form. During the in vivo (infected hamsters with *L. donovani*) activity elucidation, arbortristosides were reported with leishmanicidal activity at a concentration of 100 mg kg^{-1} for 5 days when administered orally and 10 mg kg^{-1} for 5 days when administered intraperitoneally [25].

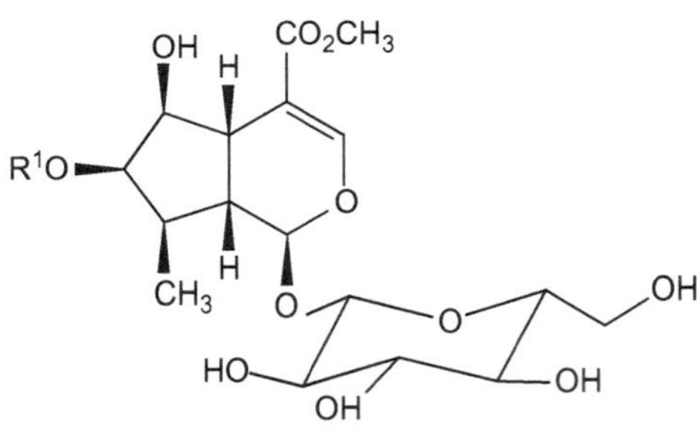

Picroside I (R^1 =Vanilloyl, R^2 =H) and kutkoside (R^1 =H, R^2 =Cinnamoyl): Rhizomes of *Picrorhiza kurroa* are a good source of both compounds that were reported to increase the nonspecific immune response and offer a high degree of protection against the *L. donovani* promastigotes in hamsters [26].

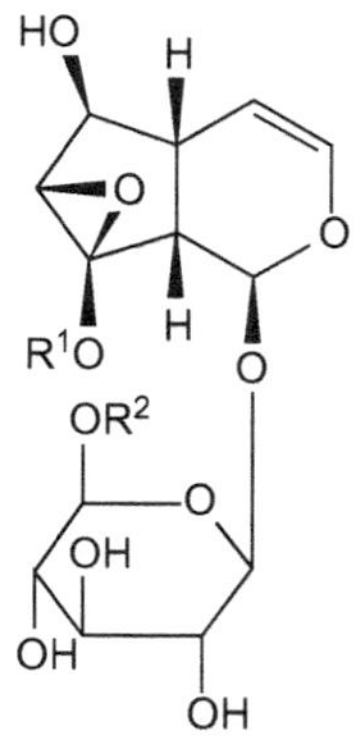

Amarogentin: It is a secoiridoid glycoside extracted from *Swertia chirata* (Gentiaceae). It inhibited the catalytic activity of the topoisomerase I from *L. donovani* when evaluated at a concentration higher than 60 μM. Therefore, it plays an important role in disturbing the cellular functions of the *Leishmania* parasite because Topoisomerase I catalyses the relaxation of DNA [27].

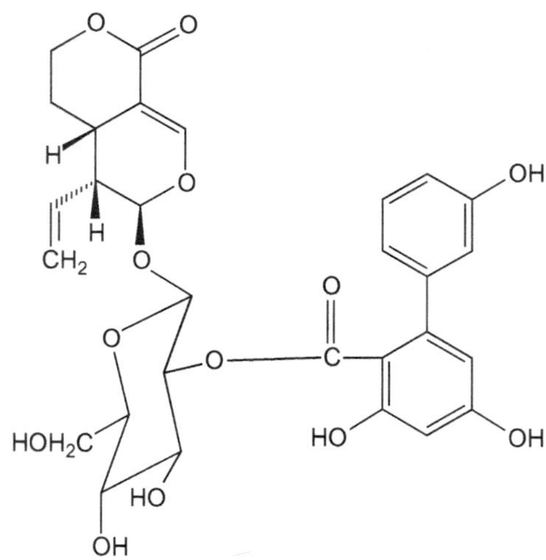

Grifolin: It is a monoterpenoid isolated from *Peperomia galoides* that caused the total lysis of the *L. donovani* promastigote at a concentration of 100 μg mL^{-1} [28].

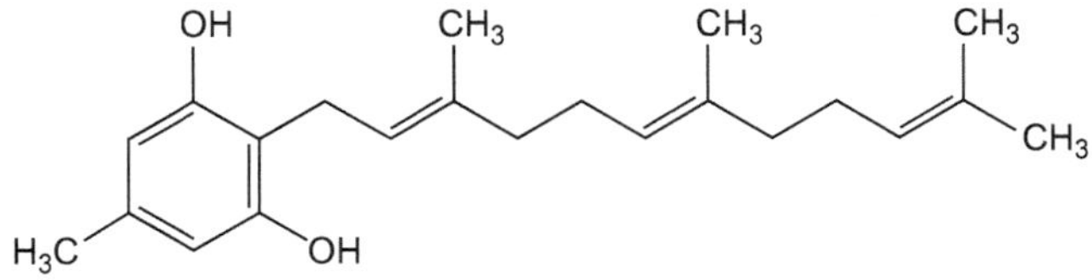

Piperogalin: *Peperomia galoides* is also a source of monoterpenoid piperogalin. Mahiou et al. [28] reported that it causes the lysis of more than 90% of *L. donovani* promastigotes.

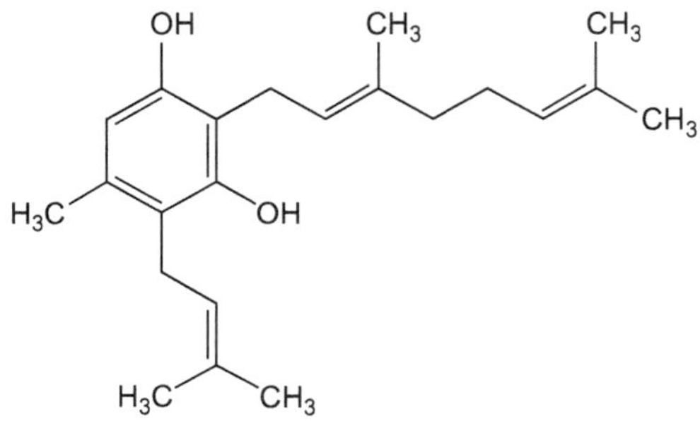

Kudtriol: It is a sesquiterpene alcohol from the aerial parts of the plant *J. glutinosa*. It has shown toxic effects against *L. donovani* promastigotes at a concentration of 250 μg mL^{-1} [29].

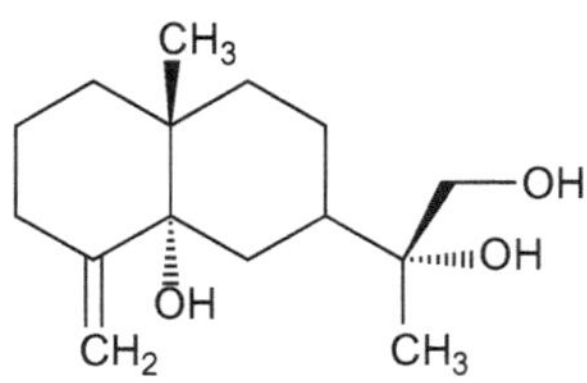

Jatrophone and Jatrogrossidione: These diterpenes possess high toxic activity against the promastigote forms of the VL-causing agent L. *chagasi*. The antileishmanial activity of Jatrogrossidione is greater than that of Jatrophone (IC**100** = 0.75 and 5 μg mL^{-1}) [30].

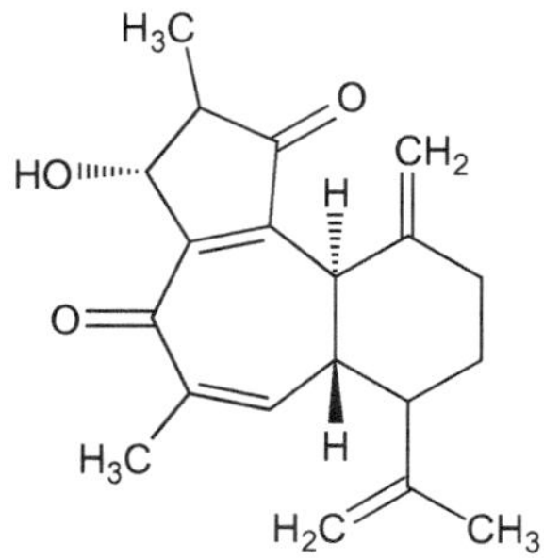

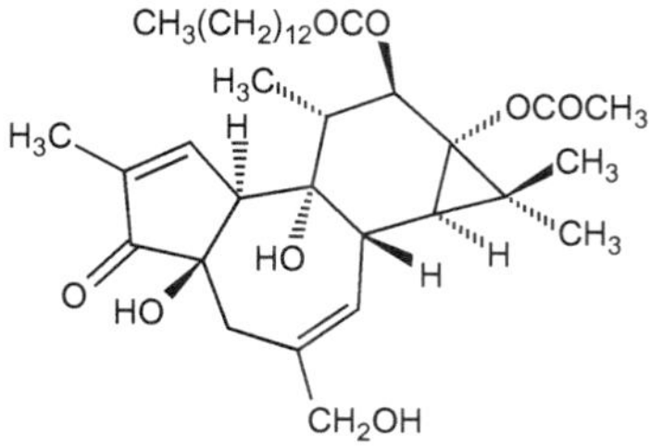

Ribenol: It is an *ent*-manoyl oxide derivative extracted from *Sideritis varoi* (Lamiaceae). It has been reported to show in vitro antileishmanial activity against *L. donovani* promastigotes [31].

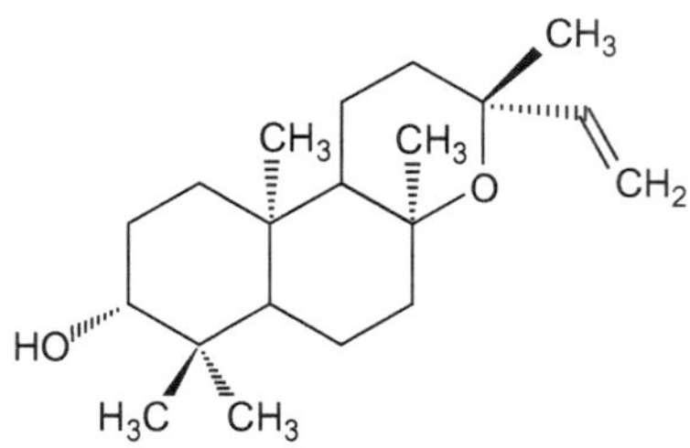

6-β-hydroxyrosenonolactone: It is also a diterpene isolated from the bark of *Holarrhena floribunda* (Apocynaceae). It has sensible activity against the promastigotes of *L. donovani* but weak leishmanicidal activity against its amastigote form [32].

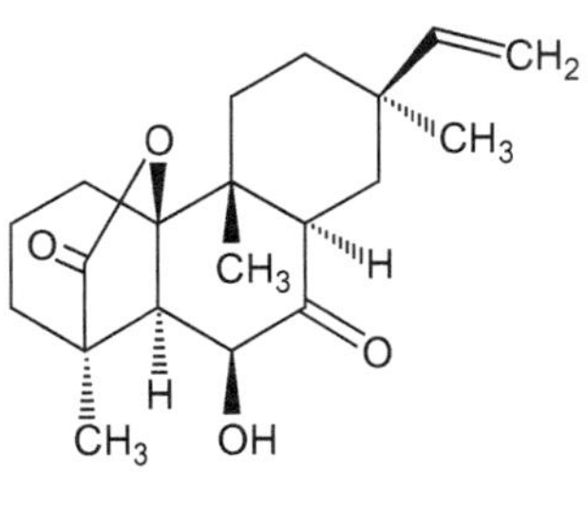

(24Z)-3-oxotirucalla-7,24-dien-26-oic acid and *epi*-oleanolic acid: They are two triterpenes isolated from the leaves of *Celaenododendron mexicanum* (Euphorbiaceae). They exhibited leishmanicidal activity on *L. donovani* promastigotes (IC**50** = 13.7 and 18.8 μM, respectively) [18].

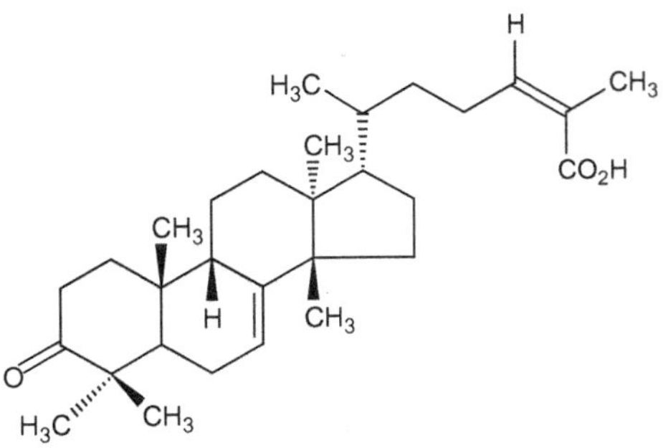

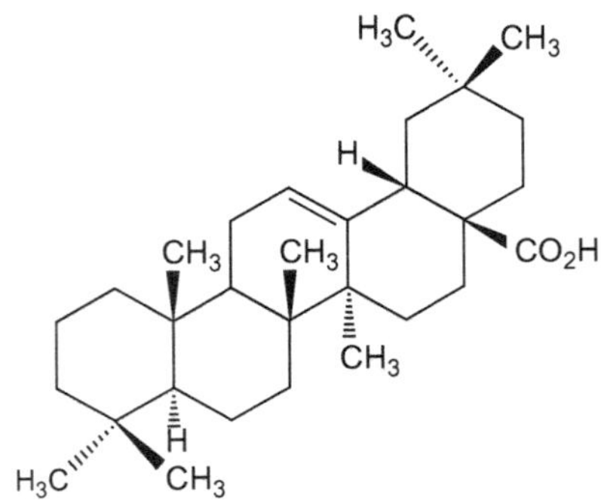

Simalikalactone D and 15-β-heptylchaparrinone: They are triterpenes that show activity against *L. donovani* promastigotes [33].

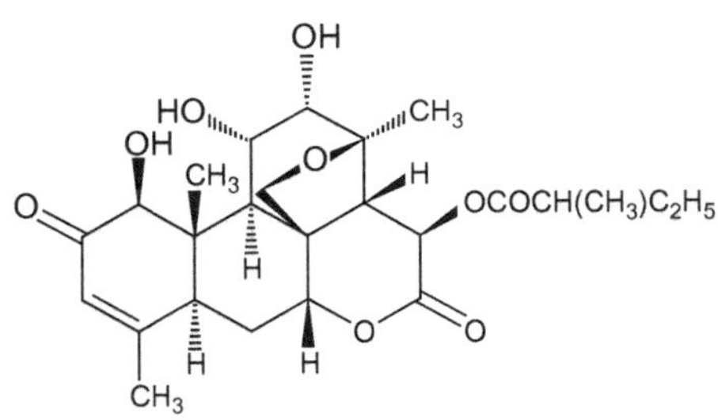

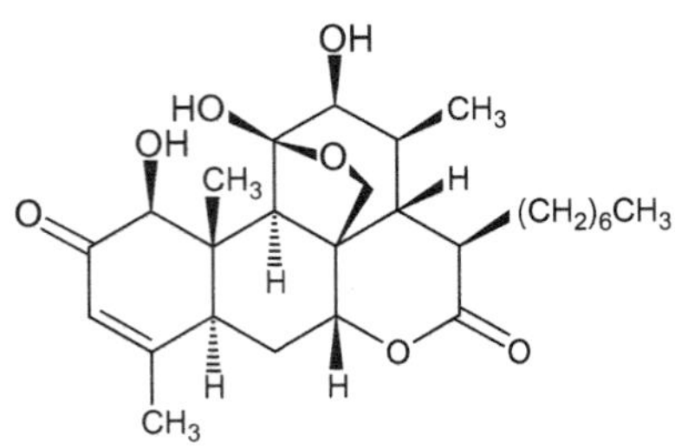

Mimengoside A: It is a saponin isolated from *Buddleja madagascariensis* (Loganiaceae). It has toxic activity against *L. infantum* promastigotes [34].

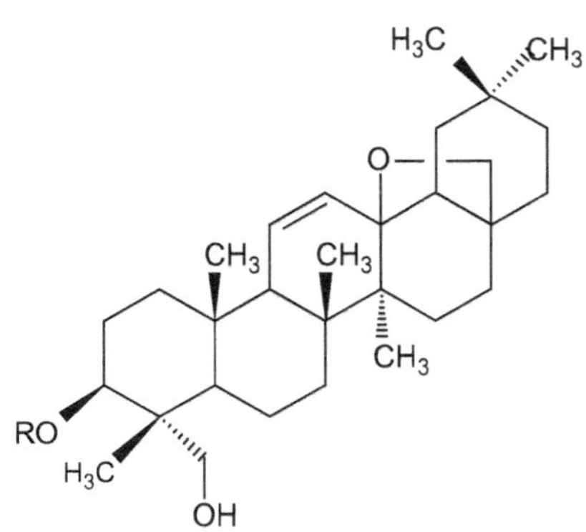

Chalcone: The chalcone (*E*)-1-[2,4-hydroxy-3-(3-methylbut-2-enyl)phenyl]-3-[4-hydroxy-3-(3-methylbut-2-enyl)phenyl]-prop-2-en-1-one 73 is isolated from *Piper aduncum* (Piperaceae). It was reported to be toxic to the promastigotes of *L. donovani* [35].

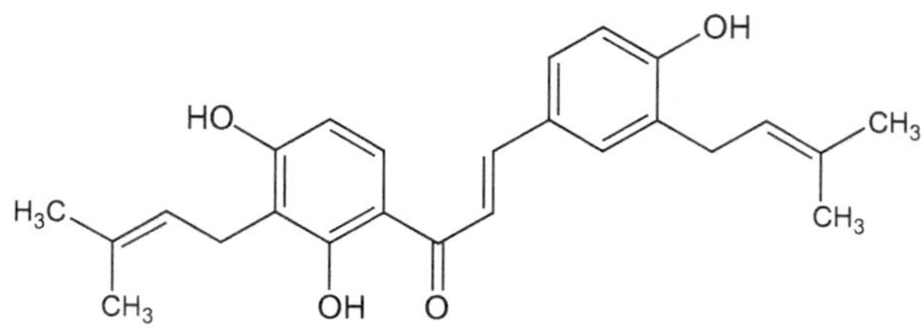

Licochalcone A: This saponin is isolated from the roots of the Chinese licorice plant (*Glycyrrhiza*spp., Fabaceae), and it inhibited the in vitro growth of *L. donovani* promastigotes [36].

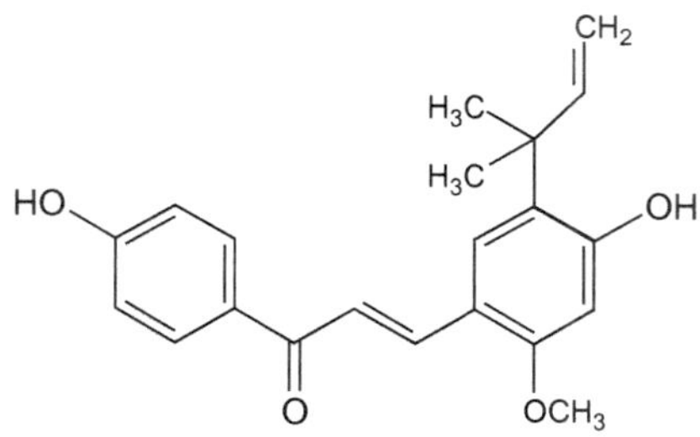

6-hydroxybenzofuran-3(2*H*)-one: It is an aurone with activity against *L. donovani* promastigotes (EC**50** = 0.09–0.11 μg mL^{-1}) and amastigotes (EC**50** = 1.24 μg mL^{-1}) [37].

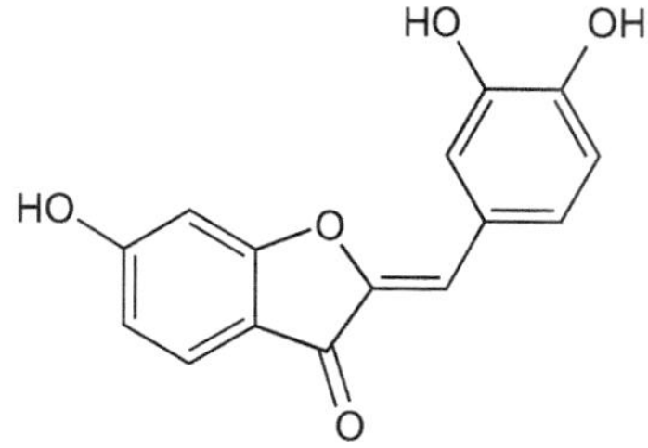

Podocarpusflavone (R1 = H, R2 = 0CH$_3$): It is extracted from the leaves of *Celanodendron mexicanum*. Antileishmanial activity against *L. donovani* promastigotes was reported [18].

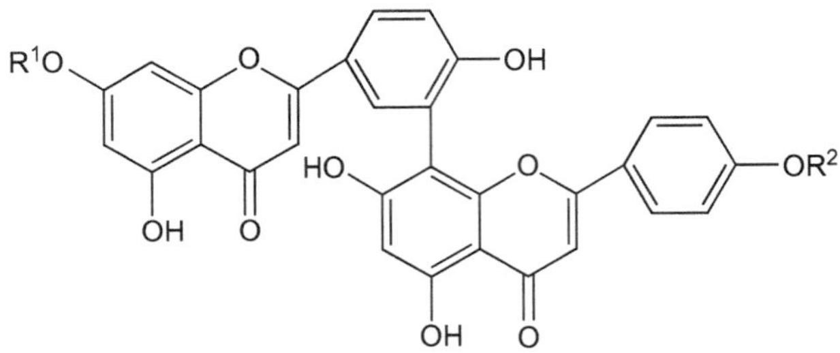

Senegalene: It belongs to acetogenins isolated from the seeds of *Annona senegalensis* (Annonaceae). Antileishmanial activity was reported against *L. donovani* promastigotes at varying concentrations between 25 and 100 μg mL^{-1} [38].

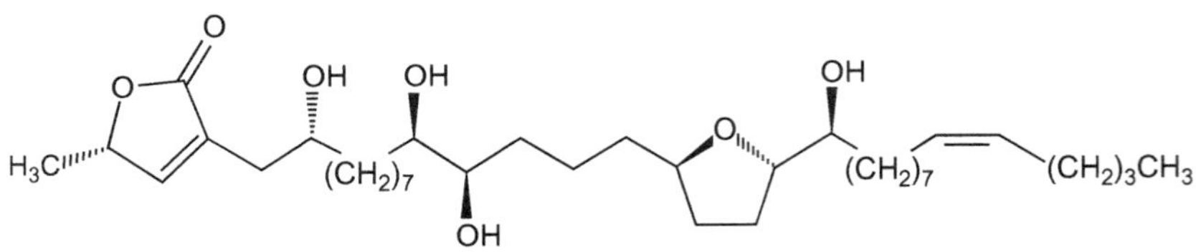

Rolliniastatin-1: It is another acetogenin from the stem bark of *Rollinia emarginata* (Annonaceae); antileishmanial activity was reported against the promastigote forms of L. *donovani* [39].

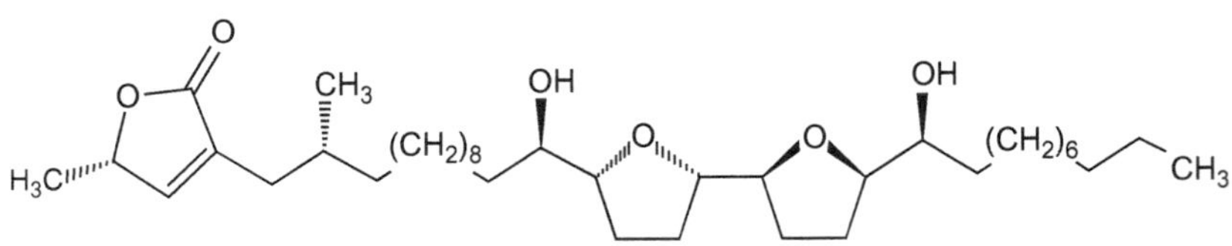

Argentilactone: It is an α,β-unsaturated δ-lactone isolated from *Annona haemantantha* (Annonaceae) roots. Significant activity was reported against *L. donovani* promastigotes [40].

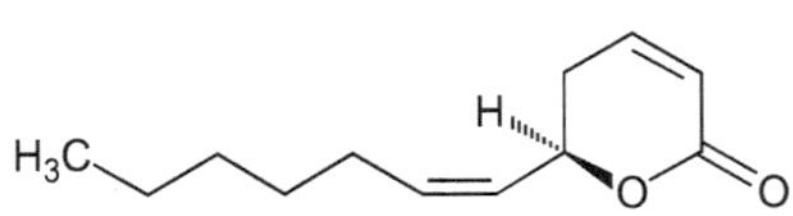

Camptothecin: It is a pentacyclic alkaloid isolated from *Campthotheca acuminata* Decsn. (Nyssaceae). Its analogue was reported to have powerful antileishmanial activity against *L. donovani* promastigote [41].

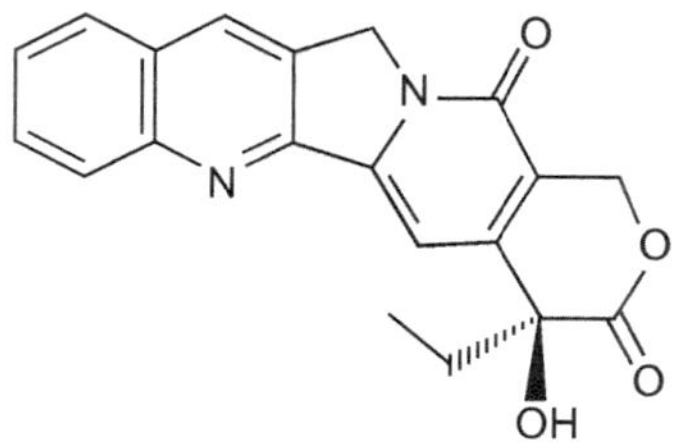

Ancistroealaine A: It is isolated from *Ancistrocladus ealaensis* and it was more active against amastigotes of *L. donovani* than pentostam [42].

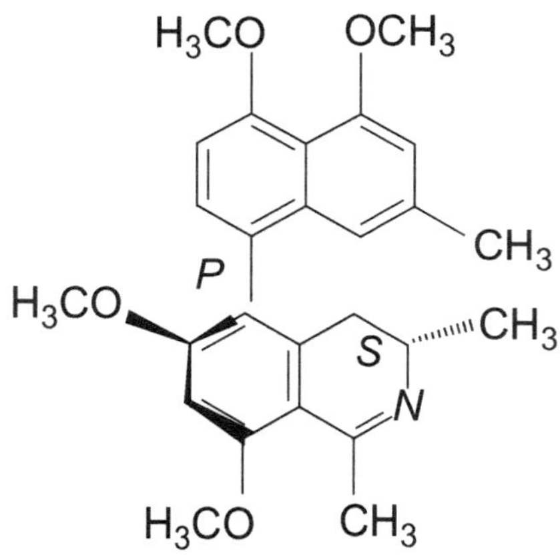

***N*-methylliriodendronine:** It was extracted from *Stephania dinklagei* Diels (Menispermacae) and was reported to be most active against amastigotes of *L. donovani* (IC50 = 36.1 mM) [43].

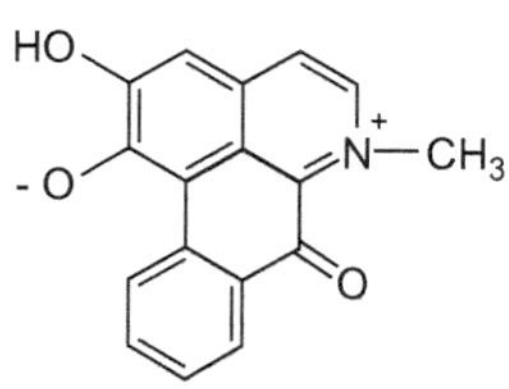

Klaivanolide: It is an extract from the stems of *Uvaria klaineana* (Annonaceae). It is a novel bisunsatured seven-membered lactone (5-acetoxy-7-benzoyloxymethyl-7Hoxepin-2-one). It showed potent in vitro antileishmanial activity against both sensitive and amphotericin B-resistant L. *donovani* promastigotes with IC50 values of 1.75 and 3.12 mM 88, respectively [44].

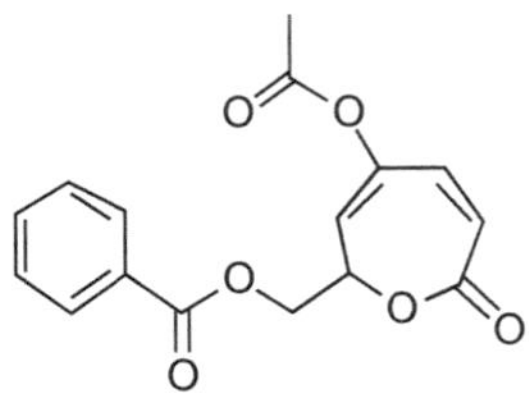

Licochalcone A: It is a chalcone isolated from *Piper aduncum* L. (Piperaceae). It was active against the intracellular amastigote of L. *donovani* with cytotoxicity varying between 0.19 and 2.06 mg/mL and EC50 values between 0.39 and 0.41 mg/mL [45].

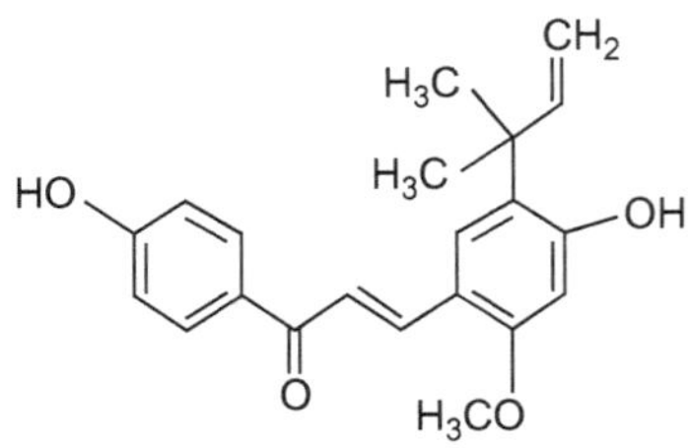

Problems associated with antiparasitic natural product research

There are many challenges in developing a natural product such as an antiparasitic molecule.

1. The quality and uniformity of the chemical constituents and their concentrations in natural products may be affected by factors such as the subspecies or age of the plant, the geographical and seasonal variation, the time and method of collection and storage, and other environmental variables. There may be major difficulties in isolating the pure compounds from complex mixtures of substances often found in extracts submitted for screening.
2. There is a high probability that many of the products that have been classified as active against a particular organism in vitro are likely to be general toxins and quite unsuitable for development as an antiparasitic product.
3. The lack of correlation between tests in vitro and in vivo where we often see good activity in vitro but poor or no activity in vivo.
4. Converting the natural product to a metabolically robust, orally bioavailable drug is extremely challenging, and this takes a very long time.

Despite these issues, natural products have highly significant roles in drug discovery. The renaissance in VL or *Leishmania*/HIV coinfection lacks effective therapies. Therefore, more efforts in the discovery of new specific and safe drugs from natural resources are needed. The literature reported that a wide variety of plant extracts as natural products have many active compounds that have shown very good activity against VL-causing parasites. The different molecules that have been discussed above show a broad range of structural diversity, which could constitute a useful source for new antileishmanial compounds for new treatments. The further necessity is to amplify studies on potent molecules that have good activity both in vitro and in vivo. There is need for finding new candidate molecules and in-depth study from natural products because it could be safe, effective, and low-cost pharmaceuticals and could be panacea for VL. The diverse chemical constituents derived from natural products should be accelerated, which would be relevant to the future of VL drug discovery.

References

[1] R. Lira, S. Sundar, A. Makharia, R. Kenney, A. Gam, E. Saraiva, D. Sacks, Evidence that the high incidence of treatment failures in Indian kala-azar is due to the emergence of antimony-resistant strains of *Leishmania donovani*, J Infect Dis 180 (1999) 564–567.

[2] R. Deb Roy, Malarial Subjects: Empire, Medicine and Nonhumans in British India, 1820–1909, Cambridge (UK): Cambridge University Press, 2017. Chapter 3, 'A Cinchona Disease': Making Burdwan Fever. Available from: https://www.ncbi.nlm.nih.gov/books/NBK481427/.

[3] WHO, Overcoming Antimicrobial Resistance: World Health Report on Infectious Diseases 2000, World Health Organization, Geneva, 2000. Available at: http://www.who.int/infectious-disease-report/2000/ch4.htm.

[4] S. Sundar, K. Pai, R. Kumar, K.P. Tripathi, A.A. Gam, M. Ray, R.T. Kenney, Resistance to treatment in kala-azar: speciation of isolates from northeast India, Am. J. Trop. Med. Hyg. 65 (3) (2001) 193–196.

[5] M. Grogl, A.M. Oduola, L.D. Cordero, D.E. Kyle, *Leishmania* spp.: development of pentostam-resistant clones in vitro by discontinuous drug exposure, Exp. Parasitol. 69 (1989) 78–90.

[6] J. Alexander, D.G. Russell, The interaction of Leishmania species with macrophages, Advances in Parasitology Academic Press 31 (1992) 176.

[7] S. Gupta, Nishi, Visceral leishmaniasis: Experimental models for drug discovery, Indian J. Med. Res. 133 (2011) 27–39.

[8] A. Zeevi, The immune system: therapeutic targets and monitoring tests, in: Presented at the 37th American Society of Health-System Pharmacists Midyear Clinical Meeting and Exhibits; December 9, 2002; Atlanta, Georgia, 2009.

[9] R. Venkataramanan, Biopharmaceutical considerations in using new immunomodulating agents, in: Presented at the 37th American Society of Health-System Pharmacists Midyear Clinical Meeting and Exhibits; Atlanta, Georgia. December 9, 2002.

[10] A. Kumar, S.C. Pandey, M. Samant, Slow pace of antileishmanial drug development, Parasitol. Open 4 (e4) (2018) 1–11.

[11] P. Trouiller, et al., Drug development for neglected diseases: a deficient market and a public-health policy failure, Lancet 359 (2002) 2188–2194.

[12] B. Hazra, A.K. Saha, R. Ray, D.K. Roy, P. Sur, A. Banerjee, Antiprotozoal activity of diospyrin towards *Leishmania donovani* promastigotes *in vitro*, Trans. R. Soc. Trop. Med. Hyg. 81 (1987) 738.
[13] L. Maes, et al., *In vitro* and *in vivo* activities of a triterpenoid saponin extract (PX-6518) from the plant *Maesa balansae* against visceral leishmania species, Antimicrob. Agents Chemother. 48 (2004) 130–136.
[14] S.L. Croft, A.T. Evans, R.A. Neal, The activity of plumbagin and other electron carriers against Leishmania donovani and Leishmania mexicana amazonensis, Ann. Trop. Med. Parasitol. 79 (1985) 651.
[15] R.P. Borris, J.M. Schaeffer, in: H.N. Nigg, D. Seigler (Eds.), Phytochemical Resources for Medicine and Agriculture, Plenum Press, New York, 1992, pp. 117–158.
[16] A. Fournet, J.C. Gantier, A. Gautheret, L. Leysalles, M.H. Munos, J. Mayrargue, H. Moskowitz, A. Cavé, R. Hocquemiller, The activity of 2-substituted quinoline alkaloids in BALB/c mice infected with Leishmania donovani, J. Antimicrob. Chemother. **33** (1994) 537.
[17] V. Mahiou, F. Roblot, R. Hocquemiller, A. Cavé, New prenylated quinones from Peperomia galioides, J. Nat. Prod. 59 (1996) 694.
[18] M.R. Camacho, G.C. Kirby, D.C. Warhurst, S.L. Croft, J.D. Phillipson, Oxoaporphine alkaloids and quinones from Stephania dinklagei and evaluation of their antiprotozoal activities, Planta Med. **66** (2000) 478.
[19] V. Mahiou, F. Roblot, R. Hocquemiller, A. Cavé, A.R. de Arias, G. Yaluff, A. Fournet, A. Angelo, New aporphine alkaloids from guatteria foliosa, J. Nat. Prod. 57 (1994) 890.
[20] E.F. Queiroz, F. Roblot, A. Cavé, M.Q. Paulo, A. Fournet, Pessoine and spinosine, two catecholic berbines from Annona spinescens, J. Nat. Prod. 59 (1996) 438.
[21] A. Février, M.E. Ferreira, A. Fournet, G. Yaluff, A. Inchausti, A.R. de Arias, R. Hocquemiller, A. Waecher, Acetogenins and other compounds from *Rollinia emarginata* and their antiprotozoal activities, Planta Med. 65 (1) (1999) 47–49.
[22] A. Fournet, A. Angelo, V. Muñoz, R. Hocquemiller, A. Cavé, Effect of some bisbenzylisoquinoline alkaloids on American *Leishmania* sp. in BALBIc mice, Phytother. Res. 7 (1993) 281.
[23] T. Kam, K. Sim, T. Koyano, K. Komiyama, Leishmanicidal alkaloids from *Kopsia griffithii*, Phytochemistry 50 (1999) 75.
[24] T. Kam, K. Sim, T. Koyano, M. Toyoshima, M. Hayashi, K. Komiyama, Cytotoxic and leishmanicidal aminoglycosteroids and aminosteroids from Holarrhena curtisii, J. Nat. Prod. 61 (1998) 1332.
[25] J.S. Tandon, V. Srivastava, P.Y. Guru, Iridoids: a new class of leishmanicidal agents from Nyctanthes arbortristis, J. Nat. Prod. **54** (1991) 1102.
[26] A. Puri, R.P. Saxena, P.V.G. Sumanti, D.K. Kulshreshtha, K.C. Saxena, B.W. Dhawan, Immunostimulant Activity of Picroliv, the Iridoid Glycoside Fraction of Picrorhiza kurroa, and its Protective Action against Leishmania donovani Infection in Hamsters1, Planta Med. **58** (6) (1992) 528–532.
[27] S. Ray, H.K. Majumder, A.K. Chakravarty, S. Mukhopadhyay, R.R. Gil, G.A. Cordell, Amarogentin, a naturally occurring secoiridoid glycoside and a newly recognized inhibitor of topoisomerase I from Leishmania donovani, J. Nat. Prod. **59** (1996) 27.
[28] V. Mahiou, F. Roblot, R. Hocquemiller, A. Cavé, A. Angelo, A. Fournet, P. Ducrot, Piperogalin, a new prenylated diphenol from *Peperomia galioides*, J. Nat. Prod. 58 (1995) 324.

[29] L. Villaescusa-Castillo, A.M. Díaz-Lanza, M. Gasquet, F. Delmas, E. Olliver, M. Bernabé, R. Faure, R. Elias, G. Balansard, Antiprotozoal activity of sesquiterpenes from jasonia glutinosa, Pharm. Biol. **38** (2000) 176.

[30] G. Schmeda-Hirschmann, I. Razmilic, M. Sauvain, C. Morretti, V. Muñoz, E. Ruiz, E. Balanza, A. Fournet, Antiprotozoal activity of Jatrogrossidione from *Jatropha grossidentata* and Jatrophone from *Jatropha isabellii*, Phytother. Res. 10 (1996) 375.

[31] P. Richomme, M. Godet, F. Foussard, L. Toupet, T. Sévenet, J. Bruneton, A novel leishmanicidal labdane from Polyalthia macropoda, Planta Med. 57 (1991) 552.

[32] A. Loukaci, O. Kayser, K.U. Bindseil, K. Siems, J. Frevert, P.M. Abreu, New trichothecenes isolated from Holarrhena floribunda, J. Nat. Prod. **63** (2000) 52.

[33] C.W. Wright, J.D. Phillipson, Natural products and the development of selective antiprotozoal drugs, Phytother. Res. 4 (1990) 127.

[34] A.M. Emam, A.M. Moussa, R. Faure, A. Favel, F. Delmas, R. Elias, G. Balansard, Isolation and biological study of a Triterpenoid Saponin, Mimengoside A, from the leaves of Buddleja madagascariensis, Planta Med. **62** (1996) 92.

[35] S.B. Christensen, C. Ming, L. Andersen, U. Hjorne, C.E. Olsen, C. Cornett, T.G. Theander, A. Kharazmi, An antileishmanial chalcone from Chinese licorice roots, Planta Med. **60** (1994) 121.

[36] M. Chen, S.B. Christensen, J. Blom, E. Lemmich, L. Nadelmann, K. Fich, T.G. Theander, A. Kharazmi, Licochalcone A, a novel antiparasitic agent with potent activity against human pathogenic protozoan species of Leishmania, Antimicrob. Agents Chemother. **37** (1993) 2550.

[37] O. Kayser, A.F. Kiderlen, U. Folkens, H. Kolodziej, In vitro leishmanicidal activity of aurones, Planta Med. **65** (1999) 316.

[38] S. Shapaz, C. Bories, P.M. Loiseau, D. Cortés, R. Hocquemiller, A. Laurens, A. Cavé, Cytotoxic and antiparasitic activity from Annona senegalensis seeds, Planta Med. **60** (1994) 538.

[39] A.G. Waechter, A.I. Yaluff, A.R. de Arias, R. Hocquemiller, A. Cavé, A. Fournet, Leishmanicidal and trypanocidal activities of acetogenins isolated from *Annona glauca*, Phytother. Res. 12 (1998) 541.

[40] A.I. Waechter, M.E. Ferreira, A. Fournet, A.R. de Arias, H. Nakayama, S. Torres, R. Hocquemiller, A. Cavé, Experimental treatment of cutaneous leishmaniasis with argentilactone isolated from Annona haematantha, Planta Med. **63** (1997) 433.

[41] K.A. Werbovetz, A.K. Bhattacharjee, J.J. Brendle, J.P. Scovill, Analysis of stereoelectronic properties of camptothecin analogues in relation to biological activity, *Bioorg. Med. Chem. 8* (**2000**) 1741–1747.

[42] G. Bringmann, A. Hamm, C. Günther, M. Michel, R. Brun, V. Mudogo, Ancistroealaines A and B, two new bioactive naphthylisoquinolines, and related naphthoic acids from *Ancistrocladus ealaensis*, J. Nat. Prod. 63 (2000) 1465–1470.

[43] S. Sundar, F. Rosenkaimer, M.K. Makharkia, A.K. Goyal, A.K. Mandal, A. Voss, P. Hilgard, H.W. Murray, Trial of oral Miltefosine for visceral Leishmaniasis, Lancet 352 (1998) 1821–1823.

[44] B. Chataing, J.L. Concepcion, R. Lobaton, A. Usubillaga, Inhibition of Trypanosoma cruzi growth in vitro by *Solanum* alkaloids: a comparison with ketoconazole, Planta Med. 64 (1998) 31–36.

[45] M. Chen, L. Zhai, S.B. Christensen, T.G. Theander, A. Kharazmi, Inhibition of Fumarate Reductase in *Leishmania major* and *L. donovani* by Chalcones, Antimicrob. Agents Chemother. 45 (2001) 2023–2029.

Further reading

K.-P. Chang, D. Fong, R.S. Bray, Biology of Leishmania and leishmaniasis, in: K.-P. Chang, R. S. Bray (Eds.), Human Parasitic Diseases, vol. 1, Elsevier Science Publishers, Amsterdam, 1985, pp. 1–30.

B. Majester-Savornin, R. Elias, A.M. Diaz-Lanza, G. Balansard, M. Gasquet, F. Delmas, Saponins of the ivy plant, Hedera helix, and their leishmanicidic activity, Planta Med. **57** (1991) 260.

Antileishmanial drug screening

5

Screening of antileishmanials

Various screening methods have been developed and combined through the years. Since the infection of *L. donovani* was established in hamsters, they have been used for the routine experimental evaluation of chemotherapeutic agents against VL [1]. For screening of medicinal plants/compounds, various techniques have been developed and tried to assess their antileishmanial action [2, 3]. However, the technique described by Beveridge [4] was adopted in the present study with some modifications. It was found to be suitable, reliable, and helpful in generating additional information in a single trial. The pioneers in antileishmanial chemotherapy were Stauber and his colleagues. In 1958, they introduced the minimum time of 8 eight days for screening compounds against *L. donovani* in golden hamsters. This was the first time that the intracardiac (i.c.) route of inoculation was adopted. It was found to be the most suitable for establishing the infection as this route showed consistency in the parasite "intake" and also shortened the incubation time. In this technique, the activity of the compound was assessed on liver parasites, as most of the parasites on the day of assessment (eighth day) are in the liver. This technique is still followed in some laboratories, in spite of its many lacunae. It is well established that the drug is administered before the infection. As such, it puts more emphasis on extracellular amastigotes rather than the parasites that reside within the cells. Because the parasites live within the macrophages in an established infection, the assessment of drug efficacy on free amastigotes obviously is misconceived. Although, the time factor required with this screening methods are more, therefore, the modifications in this technique was done by Hanson et al. [2]. They have tried to overcome some of the shortcomings of Stauber's technique, suggesting that the potential antileishmanial agents should always be administered after infection is well established (3–4 weeks postinoculation). It was in this method that a spleen biopsy was incorporated to ascertain the pretreatment infection load as well as the baseline data for an individual animal. According to Beveridge's technique, the spleen biopsy of the hamster is carried out on day 25 postinfection. After 3–4 days, the drug is administered for 5 consecutive days. On day 7 of the last administration, the animals are sacrificed to assess the parasites in the spleen. A comparison is made with the baseline data of the treated animals with respect to untreated controls.

A critical appraisal of the screening techniques described above shows that none of them can provide comprehensive information about the total efficacy of a potential drug. This is because the total effect of a drug is dependent on two factors: (a) the effect of the drug on the parasites and (b) the effect on the host immune system. Many drugs are known to act through the immune machinery of the host. Similarly, drug action is greatly influenced by the host immunity [5]. In Stauber's and Hanson's techniques, the assessment is based on the effect of the potential drug on day 1–3

Visceral Leishmaniasis. https://doi.org/10.1016/B978-0-323-91124-5.00008-4

postinoculation; the same is true for the parasites of the liver. This is far from the actual situation in clinical practice where VL cases are more chronic and parasites are located in deeper organs such as the spleen and bone marrow. Further, these methods completely ignore the host immunity. Researchers advocated to treat infected animals from days 15 post inoculation. The treated and control animals are sacrificed on day 60 post inoculation. This technique can detect potential delayed action as well.

Beveridge's [4] method is more logical, as the pretreatment parasite burden is assessed by a spleen biopsy to select and pair animals carrying similar parasitic loads. However, the animals were sacrificed on day 7 posttreatment. Therefore, it is impossible to assess the belated action of therefore drug. Guru et al. [6] modified the technique where the delayed action can also be assessed on the same animal by repeated spleen biopsies. This is more rational as it gives all the information regarding the cure and survival time of the treated animals. It also allowed sufficient time to observe whether the host immunity plays any role in the confirmation of the candidate molecule as an effective drug. The path of drug discovery is very long and there are a number of steps involved in this process, including drug screenings, assays of screened molecules, *in vitro* studies, and in vivo studies. These four topics are described in the next few sections to understand the process.

Drug screening methods for antileishmanial drug discovery

Drug discovery is a highly complex and multidisciplinary process that needs a rigorous screening with a test compound. Drug screening is the process by which potential molecules are identified as drug candidates and optimized for the selection of a candidate drug to progress into clinical trials. Once screening is over, further in vitro and in vivo studies are required to enter a molecule into a clinical trial. Broadly, there are two methods of screening compounds.

Classical methods

In this method, the phenotypic screening of a few chemical molecules (natural or synthetic) is done with cell-based assays to identify substances that have a therapeutic effect. This method is time-consuming, where a few molecules are screened/tested with a parasite to see the antieffect. However, there is also a high chance of error with this method. In the past, this approach was used for drug discovery, but nowadays quick methods such as high-throughput screening are used. Usually, this type of screening for VL is done with the log phase of *L. donovani* promastigotes (1×10^6 cells/mL) into 48-well tissue culture plates. Different concentrations of standard drugs as well as test compounds are administered to see the effect of the test compound. Untreated cells serve as the control.

High-throughput screening (HTS) methods

HTS uses the technologies of automation to collect a large amount of experimental data in a relatively short time for drug discovery. Convincing examples are on record where very potent clinically used antiparasitic drugs have failed to exert action in vitro, as their activity is either mediated through the host defense system or through metabolites [5]. Physiological factors such as bioavailability and biotransformation are totally absent in in vitro. Amastigotes are the disease-producing form of vertebrate hosts, and they reside exclusively within the macrophages of the reticuloendothelial system. The discovery that macrophages infected with amastigotes can be maintained in vitro has surpassed the use of promastigotes for screening. The macrophage-amastigote system has received more recognition as it has a close resemblance to the in vivo situation. Therefore, the choice must be the animal model. But in vivo drug testing is cumbersome and time-consuming. To make the system viable, an appropriate, quick, and easy version needs to be adopted for the rapid turnover of screening results. Such a system is known as high-throughput screening. HTS covers various methods such as the reporter gene assay, ultra-HTS (UHTS), lab on a chip, etc.

Antileishmanial drug screening with reporter gene assay

The prime prerequisite for drug testing is the availability of suitable experimental systems. Different workers have utilized promastigotes in simple media or amastigotes in macrophage culture for in vitro drug testing. For in vivo screening, various animal species have served as experimental hosts. Important among them are BALB\c mice, NMRI mice, and the hamster.

A gene that has a readily measurable phenotype that can be distinguished easily over a background of endogenous cellular proteins is known as a reporter gene. There are several genes available today that can fulfill the above criteria. The criteria for the selection of such reporters are sensitivity, dynamic range, convenience, and reliability. In general, reporter genes have the advantage of low background activity in cells, but they amplify the signal from the cell surface to produce a highly sensitive, often easily detectable response. The choice of reporter will depend on the cell line used (endogenous activity), the nature of the experiment (e.g., dynamics of gene expression versus transfection efficiency), and the adaptability of the assay to the appropriate detection method (e.g., single cell imaging versus biological screening). The aim of this review is to provide an overview of reporter genes, including their types, properties, main applications, and future prospects.

Types of reporter genes

Currently, two types of reporter gene are commercially available, the intracellular and the extracellular reporter genes. Intracellular reporter gene products are produced and retained in the cells while extracellular reporter genes encode proteins that are secreted into the culture medium. The distinctive feature of the extracellular reporter gene assay is the ability to detect reporter protein activity in the culture medium without disturbing the cells. This allows repeated experimentation and sampling of the culture medium using the same cell population. A number of reporter genes are described later and summarized in Table 5.1 that are used for screening antileishmanial drugs.

i. Green fluorescent protein (GFP)

This was discovered by Shimomura in 1962 as the companion protein to aequorin (the chemiluminescent protein from the *Aequorea* jellyfish). GFP has become an important tool in drug discovery and biological research. GFP is expressed by a variety of coelenterates, including hydrozoa (*Aequorea*, *Obelia*, and *Phialidium*) and anthozoa (*Renilla*) [11]. GFP is a monomer with an Mr. of 27,000–30,000 Da is encoded by 238 amino acids with an extremely stable conformation. The in vivo luminescence of the green light in *A. victoria* is the result of a Ca^{2+}-dependent energy transfer from the light emitter of the bioluminescent photoprotein aequorin to GFP. Wild-type GFP (wtGFP) absorbs UV and blue light with a maximum peak of absorbance at 395 nm and a minor peak at 470 nm. It emits green light maximally at 509 nm with a shoulder at 540 nm.

GFP is intrinsically fluorescent, has low toxicity, and allows easy imaging and quantification using fluorescence-activated cell sorting (FACS) or microscopy. The detection of GFP in living cells can be performed and is amenable to the real-time analysis of molecular events. The use of GFP also eliminates the need for fixation or cell permeabilization. The GFP-based assay provides a rapid, real-time assessment of changes in cellular protein expression over time, and does not require additional preparation. The GFP fluorescence activity can be detected with minimal handling and, in particular, no preparation of lysate by using a fluorescence microscope, a fluorometer, a fluorescence-activated cell sorting machine, or an imaging microplate reader. The detection of GFP fluorescence can be enhanced using a charge-coupled device (CCD) system.

ii. Luciferase

The luciferase gene has been cloned from several species of the firefly, beetle, crustacean, bacteria, and sea pansy [12]. Firefly luciferase catalyzes the oxidation of firefly luciferin, producing a short-lived flash of light that decays within a few seconds. The amount of light emitted is proportional to the activity and provides a rapid indication of the transcriptional activity of the gene. Luciferase assays can detect as little as 10^{-20} mol of enzyme, making them 100-fold more sensitive than CAT assays [7]. This gene has also allowed high-throughput screening in multiwell formats.

iii. β-Lactamase

β-Lactamases are a family of bacterial enzymes that cleave penicillins and cephalosporins. An *E. coli* form of the enzyme, TEM-1 β-lactamase, has been engineered into cytosolic, membrane-associated, or secreted forms for use in various assay formats [10]. As this is a bacterial enzyme, it has no endogenous counterpart and is an excellent choice for eukaryotic reporter gene assays. Furthermore, numerous substrates are available for colorimetric determinations of secreted enzyme activity [10, 13]. A membrane-permeable fluorogenic substrate, CCF2/AM, has also been described that enables the determination of the expression of as few as 50 β-lactamase molecules within a cell [13].

iv. β-Galactosidase

β-Galactosidase is the product of the bacterial *lacZ* gene and is widely used as an intracellular reporter gene, as it offers a number of detection options depending on which substrate is used. β-galactosidase, however, suffers from the drawback that it is an enzyme endogenous to most mammalian cell lines. Therefore, experiments must be designed to distinguish between the endogenous β-galactosidase and the exogenous bacterial enzyme. This usually requires the adjustment of the pH of the assay buffer, but this may not always be

Table 5.1 A list of reporter genes currently available for various studies.

Gene	Product	Substrate	Mode of measurement	Advantages	Disadvantages	References
Luc	Luciferase	Luciferin	Microscopy, FACS, fluorimeter	High specific activity; no endogenous activity; convenient assays	Requires substrate (luciferin) and presence of O_2 and ATP	[7]
GFP	Green fluorescent protein	No substrate required	Microscopy, FACS, fluorimeter	Autofluorescent (no substrate needed); no endogenous activity; mutants with altered spectral qualities available	Requires posttranslational modification; low sensitivity (no signal amplification)	[8]
β-*gal*	Beta-galactosidase	Lactosylceramides, lactose	Microscopy, FACS, fluorimeter	Simple colorimetric readouts; sensitive bio/ chemiluminescent assays available	Endogenous activity	[9]
β-lactamase	Beta-lactam substrates		Microscopy, FACS, fluorimeter	Catalyzes a colorimetric reaction with the substrate nitrocephin	Requires substrate	[10]

sufficient to eliminate endogenous activity, so appropriate negative controls will be needed. Several different substrates and detection formats are available [9].

When choosing the reporter gene, it is important that the product has unique enzymatic activity or is otherwise easily identified from the mixture of intra- or extracellular proteins. Vital to the success of a reporter gene assay system is the sensitivity of the detection procedure. The detection method may be based on photometry, colorimetry, radiometry, fluorescence, or immunoassay. In recent years, detection systems based on bioluminescence and chemiluminescence have become widespread because of the sensitivity and ease of detection.

Ultra-HTS (UHTS)

As the technology advances, a huge no. of compound screened at a time that is known as ultra HTS. UHTS came into use around 1998 and was associated with the advent of higher-density microplates, having 1536-well plates. It has the ability to process >100,000 samples per day. This level of screening can only be achieved using fully automated screening systems. A per day or a 66 × 1536-well plate sample can be screened with UHTS; therefore, it is known as ultra-HTS.

Lab on a chip

This is a chip-based platform using highly precise microchannels in quartz/glass with on-chip electroosmotic (electrokinetic) flow or pressure-driven microfluidics. Chips use a sipper capillary to bring compounds (~1 nL) from microplate wells onto the microchip. This allows for serial continuous flow assays inside a microchannel of a multisipper chip.

Other assays for screened molecules

Preparation of natural compounds as a test material

Water-soluble plant extracts/synthetic compounds are dissolved in the required amount of triple-distilled water [14] and then made up to the final volume with TDW. Water-insoluble plant extracts are solubilized with the minimum amount of distilled absolute alcohol and later made up to the required amount by adding TDW. The synthetic compounds were administered at the initial dose schedule of 50 mg/kg/day × 5 days, intraperitoneal (i.p.). The plant extracts can be given orally as they are present in the crude form. They need to be administered in higher concentrations (1 g/kg/day × 10 day) to locate activity in them and observe the results. Once the test compound is prepared and screened using the classical or HTS approach, more activity testing of that test compound is further done to get the hit/lead compound using the following tests/methods.

Primary assays

Immunomodulatory activities

i. **Delayed-type hypersensitivity (DTH) response (footpad swelling):** Six animals per group (control and treated) are immunized on day 0 by the i.p. administration of 0.5×10^9 sheep red blood cells (SRBCs)/rat and challenged by a subcutaneous administration of 0.025×10^9 SRBCs/mL into the right hind footpad on day +14. The test compound is administered orally from day −14 until day +13. The DTH response is measured at 24 h after the SRBC challenge on day +14 and expressed as the mean percent increase in paw volume (plethysmometrically).

ii. **Neutrophil adhesion test:** In this study, six animals per group (control and treated) are used. On the 14th day of treatment, blood samples are collected (before challenge) by puncturing the retroorbital plexus into heparanized vials. They are analyzed for total leukocyte counts (TLC) and differential leukocyte counts (DLC) by fixing blood smears and staining with Field stain I and II-Leishman's stain. After initial counts, blood samples are incubated with 80 mg/ml of nylon fibers for 15 min at 37 °C. The incubated blood samples are again analyzed for TLC and DLC. The product of TLC and the percentage of the neutrophil gives the neutrophil index (NI) of the blood sample. Percent neutrophil adhesion is calculated as shown later:

$$\text{Neutrophil adhesion } (\%) = NI_U - NI_t \cdots NI_U$$

where NI_U = the neutrophil index of the untreated blood sample. NI_t = the neutrophil index of the treated blood sample.

iii. **Hemagglutinating antibody (HA) titer:** Rats of different groups are pretreated with the test compound for 14 days. Each rat is immunized with 0.5×10^9 SRBCs/rat by the i.p. route, including the control rats. The day of immunization is referred to as day 0. The animals are treated with the test for 14 more days and blood samples are collected from each rat on day 15 for the HA titer. The titer is determined by titrating serum dilutions with SRBC (0.025×10^9 cells). The microtiter plates are incubated at room temperature for 2 h and examined visually for agglutination. The highest number dilution of serum showing hemagglutination is expressed as the HA titer.

iv. **Carbon clearance test:** This is the simplest model to test phagocytosis. It is the test of phagocytotic efficiency and can be correlated with the in vitro granulocyte test. The drug/extract is administered to mice orally or intraperitoneally after 24 h. Each mouse is given an intravenous injection of 0.3 ml of Indian ink dispersion per 30 g body weight (b.w.). Blood samples from the retroorbital venous plexus are taken at intervals of 3, 6, 9, 12, and 15 min. The carbon clearance, that is the elimination rate of carbon, from the blood is determined by turbidimetric spectroscopy at 650 nm.

In vitro assessment

Promastigote is the simplest model for screening, as the parasites multiply very well in cell-free media [15, 16]. For drug testing, the promastigotes are adjusted to a concentration of 1 to 2.0×10^6/mL of the cultivation medium. The drug is added to the experimental cultures. The inhibition of promastigote multiplication was assessed after approximately 3 days, during which the control organism multiplied 3–6 times.

Due to its simplicity, some workers prefer this method for screening [17, 18], through the promastigotes are not the target parasites. The other conditions, which make this test system of little value, are the differential metabolism and the maintenance temperature of the promastigotes [19]. Due to these problems, the use of promastigotes for drug testing has limited scope. Recently, Jackson et al. [20] developed an in vitro micro test method for drug sensitivity using the promastigotes (amastigotes converted to promastigotes in vitro). The test is quantitative, rapid, and readily applicable to parasites isolated from all major forms of human leishmaniasis. The test is conducted in a serum-free, chemically defined medium.

Amastigotes the disease, producing form of vertebrate hosts, reside exclusively within the macrophages of the reticuloendothelial system. The macrophages infected with amastigotes can be maintained in vitro which means that it has surpassed the use of promastigotes for screening. The macrophage-amastigote system has received more recognition as it has close resemblance to the in vivo situation. The system may be called the semi-in vivo system. Though this system is the most preferred in vitro screening method, the results of in vitro testing often do not correlate with those obtained in animals because of the problems of the bioavailability, biotransformation, and pharmacokinetics of the drugs. To avoid any false results, in the present work, the compounds were screened only in animal models employing well-established, documented, and accepted techniques.

In vivo assessment

In vivo assessment can be done as described earlier [21].

In vitro screening of the activity of natural compounds

Susceptibility of extracellular promastigotes to natural molecules

Logarithmic-phase promastigotes can be washed with DMEM media, adjusted to $2 \times 10^6/m^2$ in modified M199 media containing 10% heat-inactivated FBS (pH 7.2), and 50 μL of each strain plated in 96-well plate (in duplicated). Different concentrations of Ayurvedic drugs and the positive control drug SAG are prepared in modified M 199-C, added in 50 μg, and incubated for 24 h at 26 °C. The percent of killing was calculated comparing the counts per minute (cpm) at each drug concentration with the cpm in the untreated controls.

Susceptibility of intracellular amastigotes to natural drugs

Normal macrophage cell line J774 will be resuspended at $2.5 \times 10^5/\mu L$ in RPMI-1640 supplemented with 10% (volume/volume) heat-inactivated fetal calf serum (SIGMA Laboratories) and gentamycin 0.1 mL. The cells were plated in 0.4 mL on eight-chamber Lab-Tek tissue culture slides. They were allowed to adhere overnight at

37°C in a 5% CO_2 incubator. Nonadherent cells were removed, and the adherent macrophages were infected with 1×10^6 promastigotes/mL at an approximate Leishmania:macrophage ratio of 3:1 in 0.3 ml of RPMI-C. This was kept overnight at 37 °C in 5% CO_2. Free promastigotes were also removed by washing with a complete medium. The cultures were incubated overnight. At this time, different concentrations of SAG and selected ayurvedic drugs must be added to the final concentration of SAG in 10 to 1 μg/mL and ayurvedic drugs 1000, 500, 250, 125, 62.5, 31.25, and 15.62 μg/mL. First dose is given to the cultures incubated overnight and next day (in the morning) second dose of different concentration of drugs could be added and incubated overnight at 37°C in 5% CO_2 incubator and next day experiment could be terminated by fixing the slides with methanol for one minute and then stained with Giemsa stain. At least 200 cells must be counted to determine the percentage of macrophages infected and the number of parasites per infected macrophage cell. The percentage killing relative to untreated macrophages may be calculated on the basis of the comparison of total amastigotes per 100 macrophages. The percentage inhibition was calculated by the following formula [6]:

$$PI = 100 - \frac{T \times 100}{c}$$

where PI = percentage inhibition; T = number of parasites in treated samples/100 macrophage cells; and C = number of parasites in control samples/100 macrophage cells.

In vivo screening of the activity of natural compounds

Preparation and maintenance of infective inoculum in vivo

The parasites are maintained in vivo in hamsters through amastigote to amastigote passages or at times by inoculating the stationary phase promastigotes.

(i) **By amastigotes:** The hamsters with well-established infections (50–60 days old) are sacrificed. The spleen is spleen removed aseptically in a sterile Locke's solution (NaCl–4 g, KCl–0.10 g, $CaCl_2$–0.10 g, KH_2 PO_4–0.15 g, glucose–1.250 g dissolved in 500 ml of triple-distilled water, pH 7.2) and cut into small pieces. The splenic tissue is homogenized gently with the help of a motor-driven tissue homogenizer consisting of a glass mortar and a Teflon pestle. The suspension is centrifuged at a low speed of 500 rpm for 10 min at 4°C, allowing the tissue debris to settle down. The supernatant is recentrifuged under chilled conditions at 2000 rpm for 20 min. The supernatant is discarded and the sediment containing amastigotes is resuspended in Locke's solution, adjusting the count with the help of Neubaur's chamber (hemocytometer) to the concentration of 1×10^7 amastigotes/0.1 mL of suspension.

(ii) **By promastigotes:** Promastigotes harvested from 7 to 10-day-old culture are washed 2–3 times by centrifugation at 2000 rpm for 20 min (4 °C). The pellet is suspended in Locke's solution and the promastigote number is adjusted to 1×10^7/0.1 mL.

Inoculation in animals

Male golden hamsters weighing 40–45 g are inoculated with 1×10^7 amastigotes or promastigotes [3]. The animals are generally kept in plastic cages (38 × 27 × 13 cms) with a standard rodent pellet diet and water *ad libitum*.

Animals are used to develop medical treatments, determine the toxicity and safety of products, and other biomedical uses. By analyzing animal models, scientists can learn what causes disease as well as how it develops. Experimental animals are required to study disease patterns and test molecules. The next section describes animal models used in the experimental chemotherapy of VL.

Drug testing

In Stauber's and Hanson's techniques, the assessment is based on the effect of potential drug on day 1–3 post inoculation and that too on the parasites of the liver. This is far from actual situation in clinical practice where the VL cases are more chronic and parasites are located in deeper organs like spleen and bone marrow. Further these methods completely ignore the host immunity.

Mikhail and Mansour [22] advocated treatment of animals from days 15 post inoculation, the treated and control animals were sacrificed on day 60 post inoculation. This technique is capable of detecting potential delayed action as well. Beveridge's [4] method is more logical as the pretreatment parasite burden is assessed by spleen biopsy to select and pair animals carrying similar parasitic load. However, the animals were sacrificed on day 7 post treatment. It is, therefore, impossible to assess the belated action of drug, Guru et al. [6], therefore, modified the technique where delayed action can also be assessed on the same animal by repeated spleen biopsies. This is more rational as it gives all the information regarding cure and survival time of treated animals and allowed sufficient time to observe if the host immunity is to play any role.

References

[1] T.T. Pham, P.M. Loiseau, G. Barratt, Strategies for the design of orally bioavailable antileishmanial treatments, Int. J. Pharm. 454 (1) (2013) 539–552.

[2] W.L. Hanson, W.L. Chapman Jr., K.E. Kinnamon, Testing of drugs for antileishmanial activity in golden hamsters infected with L. donovani, Int. J. Parasitol. 7 (1977) 443.

[3] L.A. Stauber, E.M. Franchino, J. Grun, An eight day method for screening of compounds against Leishmania donovani in the golden hamster, J. Protozool. 5 (1958) 269.

[4] E. Beveridge, Chemotherapy of leishmaniasis, in: R.J. Schnitzer, F. Hawking (Eds.), Experimental Chemotherapy, Academic Press, I, 1963, p. 257.

[5] Katiyar, J.C., Mishra, A., Visen, P.K.S. 1986. Role of host immunity of on the efficacy of thiabendazole in rats infected with Nippostrongyylus brasiliensis Wiadomosci Parazytologiczne T., XXXII, N.R. 2-163.

[6] P.Y. Guru, A.K. Agarwal, V.K. Singha, A. Singhal, C.M. Gupta, Drug targeting in Leishmania donovani infections using tuftsin bearing liposomes as drug vehicles, FEBS Lett. 245 (1989) 204.

[7] C.E. Olesen, Y.X. Yan, B. Liu, et al., Novel methods for chemiluminescent detection of reporter enzymes, Methods Enzymol. 326 (2000) 175–202, https://doi.org/10.1016/s0076-6879(00)26055-2.
[8] R.Y. Tsein, The green fluorescent protein, Ann. Rev. Biochem. 67 (1) (1998) 509–544.
[9] D.H. Juers, B.W. Matthews, R.E. Huber, LacZ β-galactosidase: structure and function of an enzyme of historical and molecular biological importance, Protein Sci. 21 (12) (2012) 1792–1807.
[10] X. Charpentier, E. Oswald, Identification of the secretion and translocation domain of the enteropathogenic and enterohemorrhagic *Escherichia coli* effector Cif, using TEM-1 beta-lactamase as a new fluorescence-based reporter, J. Bacteriol. 186 (2004) 5486–5495.
[11] J.G. Morin, J.W. Hastings, Biochemistry of the bioluminescence of colonial hydroids and other coelenterates, J. Cell. Physiol. 77 (1971) 305–312.
[12] N.N. Ugarova, L.Y. Brovko, E.I. Dementieva, Bioluminescence and Chemiluminescence, in: D.R. Vij (Ed.), Luminescence of Solids, Springer, Boston, MA, 1998.
[13] G. Zlokarnik, P.A. Negulescu, T.E. Knapp, et al., Quantitation of transcription and clonal selection of single living cells with beta-lactamase as reporter, Science 279 (1998) 84–88.
[14] J.C. Katiyar, A. Mishra, S. Gupta, P.K.S. Visen, P.K. Murthy, A.B. Sen, Efficacy of a substituted methyl benzamidazole carbamate against developing and adult helminth parasites, Vet. Parasitol. 23 (1987) 193.
[15] J.D. Berman, in: K.P. Chang, R.S. Bray (Eds.), Experimental Chemotherapy of Leishmaniasis, A. Critical Review, Elsevier, Amsterdam, 1985, p. 283.
[16] R.A. Neal, Experimental chemotherapy of leishmaniasis-persent status and prospects, in: Nityanand, A.B. Sen (Eds.), Chemotherapy and Immunology in The Control of Malaria, Filariasis and Leishmaniasis, Tata McGraw Hill Publishing Co. Ltd., India, 1983, p. 243.
[17] J.J. Marr, R.L. Berens, D.J. Nelson, T.A. Krenitsky, J. Spector, S.W. Lafon, G.B. Elion, Antileishmanial action of 4-thiopyrazolo (3, 4-d) Pyrimidine and its ribonucleoside, Biochem. Pharmacol. 31 (2) (1982) 143.
[18] H. Seneca, J.H. Kane, J. Rockenbach, Protozoicidal properties of rimocidin, Antibiot. Chemother. 2 (1952) 435.
[19] Peters, Chemotherapy of leishmaniasis-persent status problems and prospects, in: Nityanand, A.B. Sen (Eds.), Chemotherapy and Immunology in the Control of Malaria, Filariasis and Leishmaniasis, Tata Megraw Hill Publishing Ltd., 1983, p. 243.
[20] E.J. Jackson, D.J. Tally, Y.W. Ellis, B.Y. Mebrahtu, G.P. Lawyer, J.B. Were, G.S. Reid, M.P. Daniel, B.L. Limmer, Quantitative in vitro drug potency and drug susceptibility evaluation of leishmania SP. from patients unresponsive to pentavalent antimony therapy, Am. J. Trop. Med. Hyg. 43 (5) (1990) 90.
[21] J.C. Katiyar, P.K.S. Visen, A. Mishra, S. Gupta, A.P. Bhaduri, Methyl 5 (6)-4-2-pyridyl piperazine carbonyl benzamidazole-2-carbamate a new broad spectrum antihelminthic, Acta Trop. 41 (1984) 279.
[22] J.W. Mikhail, N.S. Mansour, Mastomys albicaudatus : The African white tailed rat, an experimental host for L. donovani, J. Parasitiol. 59 (1973) 1085.

Vaccine and vaccination

6

Vaccination: History

The history of vaccination is very old. One person who experienced variolation as a child in the late 1700s was Edward Jenner, a young boy who survived the process and grew up to become a country doctor in England. As a country doctor, Jenner noticed a relationship between the equine disease known as "grease" and a bovine disease known as "cow pox." He saw that farmers who treated horses with grease lesions often saw the development of cow pox in their cows, complete with blisters similar to those seen in smallpox infection. Unlike lethal smallpox, however, the cowpox blisters eventually disappeared, leaving only a small scar at the site of each blister. At the same time, Jenner was interested when a milkmaid told him that she could not catch smallpox because she had had cowpox. Jenner noted that there were many people like the milkmaid—people who milked cows and who did not get smallpox even when exposed repeatedly. With this in mind, Jenner undertook a daring experiment in 1796: he infected a young boy with cowpox in hopes of preventing subsequent smallpox infection. After allowing the boy to recover fully from cowpox, Jenner—in an experiment that would be considered unethical by today's scientific community—intentionally infected the boy with smallpox by injecting pus from a smallpox lesion directly under his skin. As Jenner had predicted, the boy did not contract smallpox [1,2].

Although Jenner wanted to report his first case study in the Transactions of the Royal Society of London, his study was rejected. Despite this, Jenner went on to collect 23 case histories over the next months and published his own book detailing his observations. The book was called "An inquiry into the causes and effects of the variolae vaccinae, a disease discovered in some of the western counties of England, particularly Gloucestershire, and known by the name of The Cow Pox." It soon became clear that Jenner's experiments had paid off, and that intentional infection with cowpox protected people from much more serious infection with smallpox. As a result, within a few years thousands of people protected themselves from the deadly smallpox disease by intentionally infecting themselves with cowpox. Jenner's process came to be called "vaccination," after "vacca," the Latin word for cow, and the substance used to vaccinate was called a "vaccine." Now, some 200 years later, we have progressed from a time when vaccination was a rare event, and Jenner's theories about vaccination were not widely accepted, to the late 1900s when vaccines are so commonplace that most children receive multiple vaccinations before they reach their first birthdays. The result of such widespread vaccination has been a marked decrease in diseases that once ravaged the world's population. An example of this is smallpox: once a major cause of death worldwide, the smallpox virus is now found only in freezers in high-containment laboratories at the Centers for Disease Control and Prevention (CDC) in Atlanta, United States, and the Institute for Viral Preparations in Moscow, Russia.

Visceral Leishmaniasis. https://doi.org/10.1016/B978-0-323-91124-5.00001-1

A vaccine is a preparation of modified or killed organisms or components of the organisms that stimulates an immune reaction/response when given to the body. This immune reaction induces protection against attacks from that organism. The history of vaccine development spans a relatively short period of time in comparison to the history of human civilization. However, monumental advances in the field of vaccines have been made in efforts to combat infectious diseases. These advances have led to a reduction and in one case the complete eradication of the burden of some infectious diseases in the world. Throughout the history of vaccine development, milestone discoveries can be identified that have shaped the field as we know it. These milestones include the development of the first vaccine against smallpox by Edward Jenner in the 18th century. As vaccine development progresses into the 21st century, it will be important to build on the experience and knowledge generated in the past, in an effort to surpass the limitations that currently hamper the development of new and more effective vaccine technologies. Presented here is an overview of the history of vaccine development, with special emphasis on malaria. Vaccine milestones are discussed below, century-wise:

18th Century

- 1796 First vaccine for smallpox (the first vaccine ever developed).

19th Century

- 1879 First vaccine for cholera.
- 1885 First vaccine for rabies by Louis Pasteur and Émile Roux.
- 1890 First vaccine for tetanus.
- 1896 First vaccine for typhoid fever.
- 1897 First vaccine for bubonic plague.

20th Century

- 1921 First vaccine for diphtheria.
- 1921 First vaccine for tuberculosis.
- 1926 First vaccine for pertussis (whooping cough).
- 1932 First vaccine for yellow fever.
- 1937 First vaccine for typhus.
- 1945 First vaccine for influenza.
- 1952 First vaccine for polio by Jonas Salk.
- 1954 First vaccine for Japanese encephalitis.
- 1954 First vaccine for anthrax.
- 1955 Injectable polio vaccine introduced.
- 1957 First vaccine for adenovirus-4 and 7.
- 1962 First oral polio vaccine.
- 1963 First vaccine for measles.
- 1967 First vaccine for mumps.
- 1967 Smallpox eradication program started.

- 1979 Smallpox eradicated from the world.
- 1970 First vaccine for rubella.
- 1974 First vaccine for chicken pox.
- 1977 First vaccine for pneumonia (*Streptococcus pneumoniae*).
- 1978 First vaccine for meningitis (*Neisseria meningitidis*).
- 1981 First vaccine for hepatitis B (first vaccine to target a cause of cancer).
- 1985 First vaccine for *Haemophilus influenzae* type b (HIB).
- 1986 First recombinant human vaccine.
- 1992 First vaccine for hepatitis A.
- 1994 Last case of polio in the Americas.
- 1998 First vaccine for Lyme disease.
- 1998 First vaccine for rotavirus.
- 1999 Eradication of polio and measles in sight.

21st Century

- 2003 First nasal vaccine for influenza approved in the United States.
- 2006 First vaccine for human papillomavirus (a cause of cervical cancer).
- 2009 First vaccine against the H1N1 influenza virus.
- 2012 First vaccine for hepatitis E and first quadrivalent (4-strain) influenza vaccine.
- 2015 First vaccine for hand-foot-mouth disease enterovirus.
- 2015 First vaccine for dengue fever.
- 2019 First vaccine for ebola approved.

Evolution of vaccine development

English physician Edward Jenner's observation that milkmaids stricken with a disease called cowpox were rarely victims of smallpox prompted him to devise the first vaccine 200 years ago. By 1980, with the use of a modern-day vaccine, smallpox was totally eradicated apart from certain laboratory stocks. Since Jenner's time, advances in the sciences of virology, bacteriology, and immunology have led to an enhanced understanding of how the human body defends itself against invading microorganisms. The development of vaccines against more than 20 infectious diseases has revolutionized our approach to public health. Since 1980, at least 15 new or improved vaccines have become available. Today, tremendous advances in science enable scientists to devise new approaches to developing vaccines against diseases that continue to plague the world's population, such as malaria, the Epstein-Barr Virus, etc. Vaccines have been proven to be among the safest and most powerful disease-prevention tools. In general, vaccines are extremely safe. However, no vaccine is completely without adverse effects—usually minor effects like a local reaction at the injection site. Very serious adverse effects resulting from vaccination are extremely rare. Millions of children and adults have received doses of vaccines safely and effectively. Vaccines are only approved for use after stringent testing and studies because the benefits of vaccination far outweigh the health risks to the individual.

How do vaccines work?

Disease-causing organisms have at least two distinct effects on the body. The first effect is very obvious: we feel sick, exhibiting symptoms such as fever, nausea, vomiting, diarrhea, rash, and many others. Although the second effect is less obvious, it is this effect that generally leads to eventual recovery from the infection: the disease-causing organism induces an immune response in the infected host. As the response increases in strength over time, the infectious agents are slowly reduced in number until symptoms disappear and recovery is complete. How does induction of the immune response occur? The disease-causing organisms contain proteins called "antigens" that stimulate the immune response. The resulting immune response is multifold and includes the synthesis of proteins called "antibodies." These proteins bind to the disease-causing organisms and lead to their eventual destruction. In addition, "memory cells" are produced in an immune response. These are cells that remain in the bloodstream, sometimes for the life span of the host. They are ready to mount a quick protective immune response against subsequent infections with the particular disease-causing agent that induced their production. If such an infection were to occur, the memory cells would respond so quickly that the resulting immune response could inactivate the disease-causing agents, and symptoms would be prevented. This response is often so rapid that infection doesn't develop—you are immune from infection. Vaccine-induced immunity works in four ways:

1. Antibodies in the gut, airways, and genital tract, the most likely sites of infection, bind to the disease-causing bacteria or viruses, preventing them from latching onto and penetrating the cells lining these passages.
2. Antibodies in the airways, gut, genital tract, and tissues bind with and neutralize the poisons released by some disease-causing bacteria. This happens with the diphtheria toxin in the throat and the tetanus toxin in the body tissues.
3. Immune cells, called cytotoxic T cells, kill virus-infected cells and some cancer cells. The measles vaccine stimulates this response.
4. Another kind of T cell helps other body cells to kill bacteria that have got inside them. They do this by releasing messenger molecules that act on the infected cells and cause inflammation. The BCG vaccine against tuberculosis (TB) does this.

Basic principles for good vaccine design

Vaccines are made in a variety of ways, depending in part on the nature of the organism and the disease it causes. Animal viruses, weakened microbes, killed microbes, and toxins are the most common components of the vaccines that have been used over the centuries. The basic principles for the good design of a vaccine are:

1. Induces an immune response that prevents disease (not necessarily infection).
2. Long-lasting. A single dose would be optimal.
3. Induces the correct type of immune response, that is, humoral or cell-mediated.
4. Induces immune response in the correct tissue location that is, gut or lungs.
5. Includes the correct antigens that stimulate the immune response to the disease or infection.

6. Delivery at the optimum time of life to prevent disease.
7. Safe with no side effects.
8. Stable (preferably long-lasting on the shelf at ambient temperature).
9. Cheap for mass production and dosing.

How to make vaccines

Obviously, a live, virulent organism cannot be used as a vaccine because it would induce the very disease it should prevent. Therefore, the first step in making a vaccine is to separate the two effects of disease-causing organisms. In practice, this means isolating or creating an organism, or part of one, that is unable to cause full-blown disease, but that still retains the antigens responsible for inducing the host's immune response. This can be done in many ways. One way is to kill the organism using formalin; vaccines produced in this way are called "inactivated" or "killed" vaccines. Examples of killed vaccines in common use today are the typhoid vaccine and the Salk poliomyelitis vaccine.

Another way to produce a vaccine is to use only the antigenic part of the disease-causing organism, for example, the capsule, the flagella, or part of the protein cell wall; these types of vaccines are called "acellular vaccines." An example of an acellular vaccine is the Haemophilus influenzae B (HIB) vaccine. Acellular vaccines exhibit some similarities to killed vaccines: neither killed nor acellular vaccines generally induce the strongest immune responses and may therefore require a "booster" every few years to insure their continued effectiveness. In addition, neither killed nor acellular vaccines can cause disease and are therefore considered to be safe for use in immunocompromised patients.

A third way of making a vaccine is to "attenuate" or weaken a live microorganism by aging it or altering its growth conditions. Vaccines made in this way are often the most successful vaccines, probably because they multiply in the body, thereby causing a large immune response. However, these live, attenuated vaccines also carry the greatest risk because they can mutate back to the virulent form at any time. Such a mutation would result in induction of the disease rather than protection against it. For this reason, attenuated vaccines are not recommended for use in immunocompromised patients. Examples of attenuated vaccines are those that protect against measles, mumps, and rubella. Immunity is often lifelong with attenuated vaccines and does not require booster shots.

Some vaccines are made from toxins. In these cases, the toxin is often treated with aluminum or adsorbed onto aluminum salts to decrease its harmful effects; after such treatment the toxin is called a "toxoid." Examples of toxoids are the diphtheria and tetanus vaccines. Vaccines made from toxoids often induce low-level immune responses and are therefore sometimes administered with an "adjuvant," which is an agent that increases the immune response. For example, the diphtheria and tetanus vaccines are often combined with the pertussis vaccine and administered together as a DPT immunization. The pertussis acts as an adjuvant in this vaccine. When more than one vaccine is administered together, it is called a "conjugated vaccine." Toxoid vaccines often require a booster every 10 years.

Another way of making a vaccine is to use an organism that is similar to the virulent organism but that does not cause serious disease, such as Jenner did with his use of the relatively mild cowpox virus to protect against the similar, but often lethal, smallpox virus. A more recent example of this type of vaccine is the BCG vaccine used to protect against *Mycobacterium tuberculosis*. The BCG vaccine currently in use is an attenuated strain of *Mycobacterium bovis* and requires boosters every 3–4 years.

In addition, biotechnology and genetic engineering techniques have been used to produce "subunit vaccines," which are vaccines that use only parts of an organism yet still stimulate a strong immune response. To create a subunit vaccine, researchers isolate the gene or genes that code for appropriate subunits from the genome of the infectious agent. This genetic material is placed into bacteria or yeast host cells, which then produce large quantities of subunit molecules by transcribing and translating the inserted foreign DNA. It is important to note that these subunit molecules are encoded by genetic material from the infectious agent, not from the host cell's genetic material. These "foreign" molecules can be isolated, purified, and used as a vaccine. The hepatitis B vaccine is an example of this type of vaccine. Subunit vaccines are safe for immunocompromised patients because they cannot cause the disease.

Vaccines are effective in preventing disease not only in individuals, but also in communities. This type of protection is called "herd immunity." When a disease spreads from one human to another, it requires both an infected individual to spread it and a susceptible individual to catch it. Herd immunity works by decreasing the number of susceptible people. When the number of susceptible people drops low enough, the disease will disappear from the community because there are not enough people to carry on the catch-and-infect cycle. The greater the proportion of vaccinated members of the community, the more rapidly the disease will disappear. This is the reason that school children are often required to be vaccinated before attending school. This required vaccination has resulted in the marked decrease of many once-common diseases including pertussis (whooping cough), polio, smallpox, and others. Currently, polio and measles are the targets of active worldwide eradication campaigns. A number of vaccines are now available, providing protection against one or more than one disease. Table 6.1 describes the diseases against which vaccines are available.

Edward Jenner and the development of the first safe vaccine for smallpox

Although Jenner is rightly celebrated for his development of cowpox as a safe vaccine for smallpox, he was not the first to make use of a relatively nonpathogenic virus to induce immunity. In 1774, farmer Benjamin Jesty inoculated his wife with the vaccinia virus obtained from "farmer Elford of Chittenhall, near Yetminster." In 1796, Jenner inoculated James Phipps with material obtained from a cowpox lesion. Six weeks later, he inoculated the experimental subject with smallpox without producing disease. Although this experiment justifiably lacked an appropriate control, further

Table 6.1 List of vaccine-preventable diseases.

Anthrax	Cervical cancer	Diphtheria
Hepatitis A	Hepatitis B	Haemophilus influenzae type B
Human Papillomavirus	Influenza	Japanese encephalitis
Lyme disease	Measles	Meningitis
Monkey pox	Mumps	Pertussis
Pneumococcal	Polio	Rabies
Rotavirus	Rubella	Shingles
Smallpox	Tetanus	Typhoid
Tuberculosis	Varicella	Yellow Fever

studies by Jenner established the efficacy of his vaccination procedure. For this feat, Jenner received a cash prize of 30,000 pounds and election to nearly all the learned societies throughout Europe.

Koch, Pasteur, and the germ theory of disease

In 1875, Robert Koch, a country physician with no formal scientific training, inoculated the ear of a rabbit with the blood of an animal that had died of anthrax. The rabbit died the next day. He isolated infected lymph nodes from the rabbit and showed that the bacteria contained within them could transfer disease to other animals. He developed and refined techniques necessary for the cultivation of bacteria, including the development of the agar growth medium. He was appointed to the Institute of Hygiene in Berlin, where his ultimate goal was to identify the organism responsible for the "White Death" tuberculosis.

Quite independently, Louis Pasteur began his studies of the "chicken cholera bacillus." In a serendipitous discovery, Pasteur inadvertently left a flask of the bacillus on the bench over the summer and inoculated eight chickens with this "old but viable" stock of chicken cholera bacillus. He found that not only did the chickens not die, but they did not even appear ill. Pasteur said that the virulent chicken cholera bacillus had become attenuated by sitting on the bench over the summer months. The similarity between these results and those of Jenner using vaccinia virus was immediately apparent to him. In honor of Jenner, Pasteur called his treatment vaccination. Pasteur later worked on anthrax and rabies and developed the first viable vaccine for anthrax and rabies.

Although Koch and Pasteur were contemporaries, they were intensely competitive and actually bitter enemies—of course, the outbreak of the Franco-Prussian war (1870) did nothing to cement their relationship. In a trenchant example of how not to behave toward a colleague at a scientific meeting, Koch made his way to the podium following Pasteur's lecture and said: "When I saw in the program that Monsieur Pasteur was to speak today...I attended the meeting eagerly, hoping to learn something new...I must confess that I have been disappointed, as there is nothing new in the speech which Monsieur Pasteur has just made...".

Although many consider Pasteur the "father of immunology." it is due to both his and Koch's efforts that the germ theory of disease was firmly established. Prior to this time, although the practical benefits of variolation were apparent, there was no known biological basis for either the cause of diseases or the efficacy of vaccination.

Bacille Calmette-Guérin (BCG)

Vaccination provides the most potent measure against infectious diseases. Recombinant (r)-live antigen carriers expressing defined pathogen-derived proteins represent promising candidates for future vaccination trials. Novel techniques in genome manipulation allow the construction of virulence-attenuated *Mycobacterium tuberculosis* and r-Mycobacterium bovis Bacille Calmette-Guérin (BCG) strains that could be used as homologous vaccines or as heterologous antigen delivery systems, respectively, for priming pathogen-specific immunity against infectious diseases, including tuberculosis. On the basis of recent achievements in the complete genome analysis of various target pathogens, combined with a better understanding of protective pathogen-specific immune responses, the rational design of a novel mycobacterial vaccine against a multitude of infectious diseases has become possible.

Dendritic cell-based vaccination against tumors and infectious diseases

Recent insights into the pivotal functions of dendritic cells (DCs) for the initiation and regulation of immune responses have provided the basis to design DC-based cellular vaccines for immune interventions against tumors and infectious diseases. Data collected from preclinical and clinical studies document that DC-based vaccination protocols can induce high levels of protective immunity and may exhibit immunotherapeutic potential. Although some obstacles to this novel approach must still be overcome, the development of methods to generate large numbers of DCs from precursor cells has paved the way for their clinical applications.

Transcutaneous immunization (TCI)

This is a novel immunization strategy by which the antigen and adjuvant are applied topically to induce potent antibody and cell-mediated immune responses specific for both the antigen and the adjuvant. TCI therefore combines the advantages of needle-free delivery with targeting the immunologically rich milieu of the skin. In animal studies, this simple technique induces robust systemic and mucosal immune responses against vaccine antigens. The first clinical studies confirmed that large antigens can be delivered to the skin to induce systemic immune responses, and that the adjuvant plays a critical role in the induction of robust responses in conjunction with delivery of antigens in a patch. These results provide an excellent basis for the further development of TCI for use in humans.

Vaccines for various infections of human

Some vaccines are available on the market for diseases caused by bacteria and viruses in humans. A list of bacterial diseases and vaccines is shown in Table 6.2, whereas vaccines against viral diseases are listed in Table 6.3.

Vaccine for VL

Despite research into the development of a visceral leishmaniasis (VL) vaccine, at present there is no effective human vaccine available against VL. It is very necessary to develop vaccines for protection from these devastating diseases. Chapters 12 and 13 in the later section of this book discuss the progress toward a VL vaccine. These chapters substantially describe vaccination approaches for VL and the current status of a Leishmania vaccine.

Table 6.2 Bacterial diseases and vaccines.

Bacteria	Diseases	Name of vaccine(s)
Bacillus anthracis	Anthrax	Anthrax vaccines
Borrelia burgdorferi	Lyme Disease	Lyme vaccines
Bordetella pertussis	Whooping cough	DPT vaccine
Clostridium tetani	Tetanus	DPT vaccine
Corynebacterium diphtheriae	Diphtheria	DPT vaccine
Coxiella burnetii	Q fever	Q fever vaccine
Haemophilus influenzae type B	Epiglottitis, meningitis, pneumonia	Hib vaccine
Mycobacterium tuberculosis	Tuberculosis	BCG vaccine
Neisseria meningitidis	Meningococcal meningitis	Meningococcal vaccine
Salmonella typhi	Typhoid fever	Typhoid vaccine
Streptococcus pneumoniae	Pneumococcal pneumonia	Pneumococcal conjugate & polysaccharide vaccine
Vibrio cholerae	Cholera	Cholera vaccine

Table 6.3 Vaccines against viral diseases.

Virus	Diseases	Name of vaccine(s)
Hepatitis A virus	Hepatitis A	Hepatitis A vaccine
Hepatitis B virus	Hepatitis B	Hepatitis B vaccine
Human papillomavirus	Cervical cancer, genital warts, anogenital cancers	HPV vaccine
H1N1 virus	Flu	H1N1 flu vaccines
Influenza virus	Influenza	Influenza vaccine
Japanese encephalitis virus	Japanese encephalitis	Japanese encephalitis vaccine
Measles virus	Measles	MMR vaccine, MMRV vaccine
Mumps virus	Mumps	MMR vaccine, MMRV vaccine
Polio virus	Poliomyelitis	Polio vaccine
Rabies virus	Rabies	Rabies vaccine
Rotavirus	Rotaviral gastroenteritis	Rotavirus vaccine
Rubella virus	Rubella	MMR vaccine, MMRV vaccine
Varicella zoster virus	Chickenpox, Shingles	MMR vaccine
Variola virus	Smallpox	Smallpox vaccine
Yellow fever virus	Yellow fever	Yellow fever vaccine

References

[1] P.J. Delves, The Merck Manual: Home Edition. Overview of the Immune System, 2020.
[2] The Merck Manual, Home Edition. Infections, 2020.

Different vaccination approach for vaccine development

7

Types of vaccine

There are the following types of vaccines:

(1) Inactivated vaccines
(2) Live attenuated vaccines
(3) Subunit vaccines
(4) Conjugate
(5) Toxicoid
(6) Recombinant

Inactivated vaccines

Inactivated vaccines do not contain live virus or bacteria. The disease microbes have been killed by chemicals, heat, or radiation. They cannot replicate and cause disease. Inactivated vaccines stimulate a weaker immune response than live vaccines and always require several doses to provide a protective immune response [1]. Immunity can also diminish over time and a booster shot is required to maintain immunity. Inactivated vaccines can be composed of either whole viruses or bacteria, or fractions of either.

Many early vaccines were produced this way:

1. Whole killed cell, treated with formaldehyde or heat-treated such as *Bordetella pertussis.*
2. Whole killed virus such as influenza.
3. Toxoid (purified inactive toxin) such as tetanus toxoid and diptheria toxoid.
4. Polysaccharide cell capsule such as *Heamophilus influenzae B,* meningococcus A and C.

The classic inactive vaccine (IPV) was designed by Jonas Salk in 1955 against polio. At the same time, an attenuated vaccine (OPV) was also produced by Albert Sabin.

Live attenuated vaccines

Live attenuated vaccines are made up of a living virus or bacteria that has been modified through a process to weaken (attenuate) and reduce its virulence (disease-inducing ability). These wild viruses or bacteria are attenuated in a laboratory, usually by repeated culturing. For example, the measles vaccine used today was isolated from a child with measles in 1954. Almost 10 years of serial passage on tissue culture media were required to transform the wild virus into a vaccine virus. A live attenuated vaccine creates a good immune response (cellular and antibody) and often provides life-long immunity with only one or two doses. Certain vaccines against viral diseases are

Visceral Leishmaniasis. https://doi.org/10.1016/B978-0-323-91124-5.00006-0

Table 7.1 Comparison between inactivated and live attenuated vaccines.

Feature	**Inactivated vaccine**	**Live attenuated vaccine**
Dose	High	Low
Efficacy	Poor effects; last a short time	Good effects; last a long time
Number of doses	Multiple	Single
Shelf life	Long	Short
Safety	Good	Poor
Stability	High	Low
Need for adjuvant	Yes	No
Reversion of virulence	No	Maybe
Duration of immunity	Less	Many years
Antibody response	IgA IgG	IgG

made this way, including measles, mumps, and chickenpox. Because these vaccines are live, they are contraindicated for people with impaired immunity; for example, people suffering from leukemia or HIV infection. Attenuation is performed by the following means:

- Irradiation with gamma or ultraviolet rays. Laboratory models of schistosomiasis and malaria vaccines. Parasites die after infection of the host before inducing pathology.
- Induced mutants: Microorganisms cultured *in vitro* under abnormal conditions such as high/low temperature, pH, and poor nutrients. Examples include influenza or poliomyelitis. Those that survive due to an induced mutation under these conditions will be nonvirulent when used to infect the normal host and so do not survive.
- Genetic deletion or modification: Delete the gene that confers virulence and the ability to replicate in the normal host. Examples include the *nef* gene in HIV required for replication and the deletion of the B toxin subunit in cholera that renders the vaccine nonpathogenic. However:
- Live attenuated vaccines may revert to the wild type and cause pathology.
- Immunization with a whole organism involves exposing the host to many hundreds of irrelevant antigens not required for protection.
- The ultimate aim is to immunize only with the antigens required to stimulate protection.

A comparison between live attenuated and inactivated vaccines is given in Table 7.1.

Recombinant or subunit vaccines

These vaccines don't use the entire microbe, but instead are made from only part of the microbe. Scientist uses the part that best stimulates the immune system. Because these vaccines only use a part of the microbe, the chances of adverse reactions to the vaccine are lower. Recombinant or subunit vaccine antigens are expressed in the following systems:

Prokaryote expression systems such as *E. coli*

1. Identify the antigen of interest using sera or a monoclonal antibody.
2. Select the antigen from the cDNA library and determine the sequence.

3. Select the plasmid and incorporate the antibiotic resistance gene.
4. Ligate the antigen DNA to the plasmid.
5. Transfect *E. coli* with plasmid containing the antigen gene.
6. Grow bacteria in the presence of an antibiotic.
7. Purify the expressed antigen from the bacterial lysate.

Advantages
- Well-studied system of gene expression and regulation.
- Easily manipulated.
- Rapid growth/doubling time.
- Large-scale production possible.
- Promoters allow strong expression of antigen up to 20% of bacterial protein.

Disadvantages
- Proteins often form insoluble inclusion bodies. Protein remains unfolded and therefore is poorly immunogenic, particularly important for antibody formation.
- No posttranslational modification (e.g., glycosylation). Carbohydrates are often very important antibody-inducing molecules.
- High levels of endotoxin from the bacterial cell wall. This causes the production of a nonspecific inflammatory response in the host such as TNF or IFNγ.
- This can overcome the problems of poor solubility by expressing the protein in tandem with a more soluble bacterial fusion protein.
- The expressed protein can also be tagged with another molecule so that the vaccine antigen can be purified from the bacterial lysate on an affinity column or similar.

Conjugate vaccine

These vaccines are very important, especially in infants. Some bacteria that can cause disease have a special outer coating (sugar molecules called polysaccharides) that hides them from the immune system (especially the immature immune system of infants). Conjugate vaccines link these coatings to an antigen or toxiod that an immature immune system can recognize, so it can respond and produce immunity. The linkage helps the immature immune system react to polysaccharide coatings and defend against the disease-causing bacterium.

Examples of conjugate vaccines are the *Haemophilus influenzae* type B (HIB) vaccine, the pneumococcal vaccine, and the meningococcal vaccine, in which a nontoxic tetanus toxiod or nontoxic diphtheria protein is used as the link to stimulate the immune system.

Toxiod vaccine

When the cause of an illness is the bacteria releasing a toxin that affects the body, a toxiod vaccine can provide protection. Scientists formulate a vaccine from just the deactivated toxin (called a toxiod), rather than the whole bacteria. A toxin can be inactivated by treating it with formalin, a solution of formaldehyde and sterilized water.

Toxiods are safe for use in vaccines and the immune system produces antibodies that lock onto and block the toxin. Vaccines against diphtheria and tetanus are examples of toxoid vaccines.

Recombinant vector vaccines

Vaccine antigens may also be produced by genetic engineering technology. The hepatitis B and human papillomavirus (HPV) vaccines (protects against cervical cancer) are developed this way. A live vector or naked DNA is used in this approach. Vectors are not supposed to cause disease in humans, hence they are nonpathogenic. This can result from the use of vectors that do not normally infect humans such as canarypox, or they can be attenuated forms of species that can infect humans (e.g., BCG). The general approach for a vector vaccine is:

- Select the plasmid vector and incorporate the antibiotic resistance gene.
- Ligate the vaccine antigen gene into the plasmid.
- Transfect the plasmid vector into *E.coli.*
- Grow in the presence of an antibiotic and analyze for the expression of the vaccine antigen.
- Isolate the plasmid construct and electroporate into BCG.
- Grow in the presence of an antibiotic and analyze for the expression of antigen.
- Immunize the host with BCG.
- Can incorporate many genes from different organisms into the vector in tandem. Also can insert cytokine genes (e.g., IL-18) to boost the immune response.

Advantages of live vectors such as BCG

- Gives long-term expression of antigen to the host's immune system.
- Safe for use in humans (already an approved vaccine for tuberculosis).
- Can induce both cell-mediated and antibody immune responses.
- Can incorporate many genes.
- Acts as its own natural adjuvant (BCG induces IL-12).

Disadvantages of live vectors such as BCG

- Slow growing.
- Low level of expression.
- Can reject and expel plasmid containing the vaccine antigen.
- Previous exposure of the host to *Mycobacterium* can immunize against this recombinant BCG vaccine.
- Posttranslational modification may not be optimal.

Naked DNA vaccines

- First described in 1993.
- Direct injection of plasmid DNA encoding the vaccine antigen gene.
- Host cells then express vaccine antigens.
- Imitates vaccination with a live virus.

- Can also use RNA.
- Unlike second-generation vaccines, do not need to express and purify the recombinant protein, synthesize the peptide, or use an adjuvant.

Viruses	**Influenza**
	Hepatitis B
	Herpes Simplex Virus
	HIV-1
	Rabies
	Measles
Bacteria	Mycobacterium
	Pneumococcus
	Salmonella
	Clostridium tetani
Protozoa	Plasmodium
	Toxoplasma
	Leishmania

References

[1] V. Vetter, G. Denizer, L.R. Friedland, J. Krishnan, M. Shapiro, Understanding modern-day vaccines: what you need to know, Ann. Med. 50 (2) (2018 Mar) 110–120.

Adjuvants

8

Adjuvants and leishmaniasis

Adjuvants, a Latin word meaning "to help," have been used to improve vaccine efficacy since the early 1920s [1,2]. The number and type of adjuvants have been expanded but their mechanism of action has remained largely unclear and empirical to develop novel vaccine [3]. Precisely how adjuvants augment the immune response is not known, but they appear to exert different effects to improve the immune response to vaccine antigens. As such, they: (a) improve antigen delivery to APCs, (b) promote the activation state of APCs, (c) enhance antigen processing and presentation by APCs, (d) modulate antibody avidity, and (e) stimulate CMI [4, 5]. Alum, one of the few adjuvants to be licensed for human use, was used with killed parasites; this combination showed significant resistance to subsequent infection with L. *donovani* [6]. An alum-precipitated L. *major* vaccine was used in a study in combination with BCG in a single-dose schedule against L. *donovani* in the langur model; it yielded moderate success [7].

Mycobacterium bovis BCG, known to possess immunopotentiating properties, has been used as a viable vaccine against tuberculosis since the 1920s. An intravenous inoculation of BCG was found to be both prophylactic and therapeutic in BALB/c mice against a challenge with amastigotes of L. *donovani* [8]. Mice inoculated with BCG exhibited significantly lower levels of parasite burden when compared to controls. A similar attempt in hamsters against visceral leishmaniasis (VL), however, exacerbated the disease and failed to protect the animals [6].

Some other immunopotentiators, including levamisole [9], cyclosporin A [10], *C. parvum* [11], glucan [12], derivatives of MDP [13], and saponins [14], are largely in use due to their ability to nonspecifically activate macrophages. Several other types of particulate adjuvants or delivery systems that have been used with leishmanial vaccines with variable outcomes include liposomes, microparticles, immunostimulating complexes, and micelles formed by intrinsically adjuvanted lipopeptides [15, 16]. Most of this work currently involves animal models, and the use of these systems in humans is still somewhat distant.

One of the most promising adjuvants involves the use of "nature's adjuvants," that is, soluble cytokines that are known to promote Th1 immune responses. Among these, IL-12 is essential for the induction and maintenance of Th1 immune responses by *Leishmania* vaccines, including DNA vaccines [17]. IL-12 is a key component in the early response to L. *donovani* and the neutralization of IL-12 over the first few days postinfection leads to elevated parasite burdens in both the spleen and liver [18] where as IL-12 was found to be an effective adjuvant against CL [19] and its use in a mouse model of L. *donovani* was found to be ineffective [20]. Another interesting approach to targeting the immune response has been the coupling of the potent antigen-presenting ability of dendritic cells to the delivery of proinflammatory

Visceral Leishmaniasis. https://doi.org/10.1016/B978-0-323-91124-5.00005-9

cytokines to the local site of response. For this purpose, the adoptive transfer of dendritic cells engineered by retroviral infection to secrete IL-12 has been used to augment the vaccination effect with dendritic cells pulsed with L. *donovani* antigen [21].

Certain sequences of bacterial DNA seem to have an immunostimulatory effect. Specific DNA sequences containing unmethylated dinucleotides (CpG motifs) activate B cells and dendritic cells and induce cytokine production by macrophages. The ability of CpG motifs to induce the production of IL-12 and TNF-α, which leads to a polarized Th1 response makes them particularly attractive as adjuvants for *Leishmania* vaccines. CpG-containing DNA has immunostimulatory effects in vaccination against VL [22].

Categories of adjuvants

The pragmatic approach of vaccine development involves trials of antigenic components in animals and humans with or without adjuvants. The systemic approach requires the identification, production, and purification of protective immunogens; the use of adjuvants/carriers; modes of presentation; and the determination of protective immune responses. First-generation vaccines composed of killed parasites with or without adjuvants are undergoing various stages of phase I (safety), phase II (reactivity), and phase III (efficacy) trials in humans in Brazil [23, 24]. BCG vaccines with the adjuvants have better protection and as effective as chemotherapy [25]. Aluminum salts were the first adjuvants approved by the US Food and Drug Administration for use in humans. The use of alum salts began in the 1930s, before regulatory guidelines became more stringent. More recently, approvals have been obtained in Europe for MF59 as an adjuvant component of the flu vaccine for elderly patients (Fluad, Novartis Vaccines) and AS04 (combination of alum and MPL, GlaxoSmithKline) as the adjuvant for a viral vaccine (hepatitis B, HPV). Broad types of adjuvants are:

1. **Mineral salts** such as aluminum hydroxide (“alum”), aluminum phosphate, and calcium phosphate.
2. **Oil emulsions** such as MF59 and detergent-stabilized oil-in-water emulsions.
3. **Particulate adjuvants** such as virosomes and ISCOMS (structured complex of saponins and lipids).
4. **Microbial derivatives** such as monophosphoryl lipid A, CpG motifs, and modified toxins.
5. **Plant derivatives** such as, saponins, squalene, and pristane.
6. **Endogenous immunostimulatory adjuvants** such as cytokines.

Selection of adjuvants

Developing or identifying new adjuvants is not an easy task and also there is no clear protocol or process for the identification of adjuvants. But scientists must judiciously select an adjuvant for vaccines targeting specific pathogens. They must determine the exact immune response desired for a given antigen ([26]). Adjuvant selection is greatly dependent on vaccine formulation and its application. In adjuvant selection, the most important criteria are the effectiveness and safety of the matter selected for the adjuvant. A number of characteristics are associated with successful adjuvants, but the following considerations should be kept in mind during adjuvants selection.

- Biodegradable in nature.
- Cheap in availability or production.
- Does not induce immune responses against itself in host.
- Selected compound is stable with a long shelf life.
- Promotes an appropriate immune response (i.e., cellular or antibody immunity depending on requirements for protection).

The limited immunogenicity of vaccine antigens has increased the search for and identification of adjuvants in vaccine development. Adjuvants enhance and extend the immune responses and reduce the amount of antigen required in each dose; however, few adjuvants are approved for use. Working toward the identification and selection of more effective adjuvants is now a part of a global health strategy to prepare for a possible pandemic in the future.

Mode of action of adjuvants

Only a few adjuvants (alum, AS04, MF59, AS03, AS01, and CpG ODN) are approved for use in human vaccines. Also, little information of the mechanism of action of adjuvants is available. The structural characterization of adjuvants and the identification of various pattern recognition receptors (PRRs) have enabled us to better understand the adjuvant mode of action at a molecular level. Understanding the adjuvant mode of action is critical in designing vaccines that elicit pathogen-specific effector and long-term memory responses as well as assessing the adjuvant safety at the developmental and regulatory stages.

The development of more effective adjuvants, delivery systems, and antigens can increase the immunogenicity of subunit vaccines against Leishmania parasites, modulating the balance between Th1 and Th2 cell responses and stimulating local immunity.

References

[1] V. Schijns, A. Fernández-Tejada, Ž. Barjaktarović, I. Bouzalas, J. Brimnes, S. Chernysh, S. Gizurarson, I. Gursel, Ž. Jakopin, M. Lawrenz, C. Nativi, S. Paul, G.K. Pedersen, C. Rosano, A. Ruiz-de-Angulo, B. Slütter, A. Thakur, D. Christensen, E.C. Lavelle, Modulation of immune responses using adjuvants to facilitate therapeutic vaccination, Immunol. Rev. 296 (1) (2020) 169–190, https://doi.org/10.1111/imr.12889. Epub 2020 Jun 28 *32594569*.

[2] F.E. Cox, Designer vaccines for parasitic diseases, Int. J. Parasitol. 27 (1997) 1147–1157.

[3] Petrovsky, et al., New-age vaccine adjuvants: Friend or foe?, International, BioPharm, 2007.

[4] R.L. Coffman, A. Sher, R.A. Seder, Vaccine adjuvants: putting innate immunity to work, Immunity 33 (4) (2010) 492–503.

[5] Glenn, O'Hagan, Adjuvants: progress, regress and pandemic preparedness, Expert Rev. Vaccines 6 (5) (2007) 651–652.
[6] J.C. Jarecki-Black, A.B. Glassman, T.W. Holbrook, The effect of BCG-vaccine upon experimental visceral leishmaniasis in hamsters, Ann. Clin. Lab. Sci. 14 (1984) 464–466.
[7] A. Misra, A. Dube, B. Srivastava, P. Sharma, J.K. Srivastava, J.C. Katiyar, S. Naik, Successful vaccination against Leishmania donovani infection in Indian langur using alum-precipitated autoclaved Leishmania major with BCG, Vaccine 19 (2001) 3485–3492.
[8] L.L. Smrkovski, C.L. Larson, Effect of treatment with BCG on the course of visceral leishmaniasis in BALB/c mice, Infect. Immun. 16 (1977) 249–257.
[9] H.R. Rezai, A.B. Behbehani, S. Gettner, S. Ardehali, Effect of levamisole on the course of experimental leishmaniasis in guinea-pigs and mice: haematological and immunological findings, Ann. Trop. Med. Parasitol. 82 (1988) 243–249.
[10] C. Bogdan, H. Streck, M. Rollinghoff, W. Solbach, Cyclosporin A enhances elimination of intracellular L. major parasites by murine macrophages, Clin. Exp. Immunol. 75 (1989) 141–146.
[11] J.O. Hill, Modulation of the pattern of development of experimental disseminated leishmaniasis by Corynebacterium parvum, J. Leukoc. Biol. 41 (1987) 165–169.
[12] J.A. Cook, T.W. Holbrook, Immunogenicity of soluble and particulate antigens from Leishmania donovani: effect of glucan as an adjuvant, Infect. Immun. 40 (1983) 1038–1043.
[13] K. Zehra, R. Pal, Anuradha, S.Y. Rizvi, W. Haq, B. Kundu, J.C. Katiyar, K.B. Mathur, Leishmania donovani in hamsters: stimulation of non-specific resistance by novel lipopeptides and their effect in antileishmanial therapy, Experientia 51 (1995) 725–730.
[14] W.R. Santos, I.A. Aguiar, E. Paraguai de Souza, V.M. de Lima, M. Palatnik, C.B. Palatnik-de-Sousa, Immunotherapy against murine experimental visceral leishmaniasis with the FML-vaccine, Vaccine 21 (2003) 4668–4676.
[15] N. Ali, F. Afrin, Protection of mice against visceral leishmaniasis by immunization with promastigote antigen incorporated in liposomes, J. Parasitol. 83 (1997) 70–75.
[16] A. Sjolander, T.M. Baldwin, J.M. Curtis, E. Handman, Induction of a Th1 immune response and simultaneous lack of activation of a Th2 response are required for generation of immunity to leishmaniasis, J. Immunol. 160 (1998) 3949–3957.
[17] P. Scott, G. Trinchieri, IL-12 as an adjuvant for cell-mediated immunity, Semin. Immunol. 9 (1997) 285–291.
[18] C.R. Engwerda, M.L. Murphy, S.E. Cotterell, S.C. Smelt, P.M. Kaye, Neutralization of IL-12 demonstrates the existence of discrete organ-specific phases in the control of Leishmania donovani, Eur. J. Immunol. 28 (1998) 669–680.
[19] F.P. Heinzel, D.S. Schoenhaut, R.M. Rerko, L.E. Rosser, M.K. Gately, Recombinant interleukin 12 cures mice infected with Leishmania major, J. Exp. Med. 177 (1993) 1505–1509.
[20] P.C. Melby, A. Tabares, B.I. Restrepo, A.E. Cardona, H.S. McGuff, J.M. Teale, Leishmania donovani: evolution and architecture of the splenic cellular immune response related to control of infection, Exp. Parasitol. 99 (2001) 17–25.
[21] S.S. Ahuja, R.L. Reddick, N. Sato, E. Montalbo, V. Kostecki, W. Zhao, M.J. Dolan, P.C. Melby, S.K. Ahuja, Dendritic cell (DC)-based anti-infective strategies: DCs engineered to secrete IL-12 are a potent vaccine in a murine model of an intracellular infection, J. Immunol. 163 (1999) 3890–3897.
[22] M. Ghosh, T. Mukherjee, Stage-specific development of a novel adenosine transporter in Leishmania donovani amastigotes, Mol. Biochem. Parasitol. 108 (2000) 93–99.
[23] K. Bahar, Y. Dowlati, B. Shidani, M.H. Alimohammadian, A. Khamesipour, S. Ehsasi, R. Hashemi-Fesharki, S. Ale-Agha, F. Modabber, Comparative safety

and immunogenicity trial of two killed Leishmania major vaccines with or without BCG in human volunteers, Clin. Dermatol. 14 (1996) 489–495.
[24] J. Convit, M. Ulrich, N. Aranzazu, P.L. Castellanos, M.E. Pinardi, O. Reyes, The development of a vaccination model using two microorganisms and its application in leprosy and leishmaniasis, Lepr. Rev. 57 (Suppl 2) (1986) 263–273.
[25] J. Convit, P.L. Castellanos, A. Rondon, M.E. Pinardi, M. Ulrich, M. Castes, B. Bloom, L. Garcia, Immunotherapy versus chemotherapy in localised cutaneous leishmaniasis, Lancet 1 (1987) 401–405.
[26] F.R. Vogel, M.F. Powell, A compendium of vaccine adjuvants and excipients, Pharm. Biotechnol. 6 (1995) 141–228, https://doi.org/10.1007/978-1-4615-1823-5_7.

Vaccine delivery system

9

Delivery systems of vaccines

For better result, some promising delivery systems were developed and they are now used for the delivery of vaccines [1, 2]. Some vaccine delivery systems are described below:

1. **Virosomes**: These have been developed by incorporating the hemagglutinin from the influenza virus into liposomes. This influenza virus surface glycoprotein guides the virosomes to antigen-presenting cells (APCs) and leads to fusion with their endosomal membrane. This process provides the optimal processing and presentation of the antigens to immunocompetent cells and results in the elicitation of humoral and cellular immune responses.
2. **Plasmid DNA**: In nucleic acid-based vaccination, plasmid DNA containing antigen-encoding sequences is delivered in a way that supports the in vivo expression and immunogenic presentation of the protein. DNA vaccines are attractive candidates for the specific immunotherapy of extra- and intracellular pathogens because they prime the neutralizing antibody and cytotoxic T cell responses. Antigen expressed from a DNA vaccine adopts its correct three-dimensional conformation (or oligomerization), ensuring the integrity of conformational epitopes binding neutralizing antibodies. DNA vaccines stimulate T cell responses to peptides generated in endogenous or exogenous processing pathways (without interference by viral proteins). Currently, DNA vaccines are developed that codeliver intrinsic adjuvants to enhance and/or modulate the immune response.
3. **Bacterial ghost system**: They are nonliving bacterial cells devoid of cytoplasmic content that maintain their cellular morphology and native surface antigenic structures, including bioadhesive properties. They are produced by the PhiX174 protein E-mediated lysis of Gram-negative bacteria. The retention by ghosts of the morphological characteristics and structural integrity of their living counterparts makes them attractive for use as vaccines. The extended bacterial ghost system is an alternative to living bacterial delivery systems and may have an added advantage because of its safety and flexibility. The endotoxin component of the Gram-negative outer membrane does not limit the use of ghosts as vaccine candidates because of the minimal toxicity of the cell-associated lipopolysaccharide compared to the free soluble form.
4. **Attenuated strains**: Attenuated strains of *Salmonella enterica serovar Typhi* (*S. typhi*) or *Typhimurium (S. typhimurium*) and *Shigella* can serve as mucosally administered live vectors that express foreign protein or polysaccharide antigens and deliver them to the immune system, eliciting protective immune responses. Attenuated strains of salmonella and shigella can also carry plasmids with foreign genes under the control of the eukaryotic expression systems (DNA vaccines) and, following mucosal administration, can deliver those plasmids to APCs.
5. **Intracellular bacterial vectors**: They infect phagocytic cells, aiding in the delivery of antigens directly to professional APCs, which express costimulatory molecules and stimulate the efficient priming of a T cell response. Recently, the use of three intracellular bacterial vectors, *Listeria monocytogenes, Salmonella typhimurium,* and the BCG vaccine strain of *Mycobacterium bovis,* has become more common.

Visceral Leishmaniasis. https://doi.org/10.1016/B978-0-323-91124-5.00010-2

6. **Red blood cells (RBC)**: It can be used as an antigen delivery system for the enhancement of immunotargeting for vaccination. The method for coupling antigen on the autologous RBC membrane is based on biotin-avidin-biotin bridges. The use of RBC-based immunization strategies does not require adjuvants. RBC-based immunization does not have any side effects and at the mean time assures a prolonged exposure to the antigen. The appreciable titer of antiavidin antibodies could be advantageous for the opsonization of antigen-carrying RBCs, facilitating their recognition by macrophages. Furthermore, the antigen-carrying RBCs appear to be an effective tool for antigen delivery into dendritic cells, and INF-g could be used advantageously to augment the ability of RBC to induce Th1 responses.
7. **Phospholipid microspheres**: This is a novel delivery system prepared from polylactic-coglycolic acid (PLGA) and phosphatidyl ethanol amine in the molar ratio of 1:71 to deliver drugs to macrophages in experimental leishmaniasis. Mannose-grafted microspheres may have possible application in clinics, not only in visceral leishmaniasis but also in other macrophage-associated disorders.
8. **Liposomes**: They are well-known drug carrier systems made from phosphatidylethanolamine, cholesterol, and dicetyl phosphate in the molar ratio of 7:2:1. They were used as the control to compare the efficacies of the two carriers. The presence of mannosyl fucosyl receptors on macrophages intercalated in both mannose-grafted liposomes and mannose-grafted microspheres. Plain arsonoliposomes and phosphatidylcholine arsonoliposomes prepared with palmitoyl-(C16) or lauroyl-(C12) acyl side chain arsonolipids showed in vitro antileishmanial activity after a 72 h incubation period against the wild-type promastigote forms of *Leishmania donovani*. Arsonolipids dissolved in dimethyl sulfoxide were found not to be potent against malaria.

The application of new strategies to develop effective delivery vehicles is essential in modern vaccine design. Important scientific, cultural, temporal, and secular issues impact the development and delivery of vaccines [3]. In this chapter, we discuss the impact of demographics, regulatory science, the antivaccine movement, and finally the impact of new biology and individualized medicine, which we call vaccinomics, on vaccine development and delivery.

References

[1] E.L. Giudice, J.D. Campbell, Needle-free vaccine delivery, Adv. Drug Deliv. Rev. 58 (1) (2006) 68–89.
[2] C. Saroja, P. Lakshmi, S. Bhaskaran, Recent trends in vaccine delivery systems: a review, Int. J. Pharm. Investig. 1 (2) (2011) 64–74, https://doi.org/10.4103/2230-973X.82384.
[3] G.A. Poland, R.M. Jacobson, I.G. Ovsyannikova, Trends affecting the future of vaccine development and delivery: the role of demographics, regulatory science, the anti-vaccine movement, and vaccinomics, Vaccine 27 (25–26) (2009) 3240–3244.

Routes of vaccination

10

Vaccination route

The vaccination route is recommended after the development of a vaccine. (https://www.cdc.gov/vaccines/hcp/acip-recs/general-recs/administration.html). Deviation from the recommended route may reduce vaccine efficacy or increase local adverse reactions. Many routes of vaccine delivery have been explored to date [1]. They are described below:

1. **Intranasal route (NAS)**: The live attenuated influenza vaccine (LAIV, FluMist) is currently the only vaccine administered by the nasal route. The vaccine dose (0.2 mL) is inside a special sprayer device. A plastic clip on the plunger divides the dose into two equal parts. The patient should be seated in an upright position with the head tilted back. Instruct the patient to breathe normally. The provider should gently place a hand behind the patient's head. The tip of the nasal sprayer should be inserted slightly into the naris. Half the contents of the sprayer (0.1 mL) are sprayed into the nostril. The dose-divider clip is then removed, and the procedure is repeated in the other naris. Detailed information on the nasal administration of LAIV is included in the manufacturer's product information. The dose does not need to be repeated if the patient coughs, sneezes, or expels the dose in any other way [2].

   It is possible for the LAIV spray to cause low-level contamination of the environment with the vaccine virus, but there have been no reports of vaccine virus transmission by this route. No instances have been reported of illness or attenuated vaccine virus infections among inadvertently exposed healthcare personnel or immunocompromised patients. Only healthcare personnel with severe immunosuppression (i.e., who require a protective environment, such as for hematopoietic cell transplant) should not administer LAIV. Healthcare personnel with this level of immunosuppression would not be administering vaccines. Other healthcare personnel at higher risk for influenza complications can administer LAIV. These include persons with underlying medical conditions placing them at higher risk or who are likely to be at risk, including pregnant women, persons with asthma, and persons aged 50 years and older.
2. **Oral route (PO)**: Rotavirus vaccines (RV1/Rotarix, RV5/RotaTeq) and oral typhoid (TY21a/Vivotif) are the only vaccines licensed in the United States that are administered by the oral route. RV1/Rotarix requires reconstitution prior to oral administration. Oral vaccines should generally be administered prior to administering injections or performing other procedures that might cause discomfort. Administer the liquid slowly down one side of the inside of the cheek (between the cheek and gum) toward the back of the infant's mouth. Care should be taken not to go far enough back to initiate the gag reflex. Never administer or spray (squirt) the vaccine directly into the throat. Detailed information on the oral delivery of these vaccines is included in each manufacturer's product information (https://www.who.int/ith/vaccines/rotavirus/en/). ACIP does not recommend readministering a dose of rotavirus vaccine to an infant who regurgitates, spits out, or vomits during or after administration. No data exist on the benefits or risks associated with readministering a dose. The infant should receive the remaining recommended doses of rotavirus vaccine following the routine schedule (with a 4-week minimum interval between doses). There are no restrictions on the

Visceral Leishmaniasis. https://doi.org/10.1016/B978-0-323-91124-5.00009-6

infant's consumption of breast milk or any other liquid before or after the administration of either of these vaccines.

3. **Intradermal route (ID)**: Fluzone Intradermal is the only vaccine licensed in the United States that is administered by the intradermal route. It is approved only for use in persons 18 through 64 years of age. This Fluzone formulation is not the same as the intramuscular formulations of inactivated influenza vaccine (TIV). Other TIV formulations should NOT be administered by the intradermal route [3].

   Site: The site of administration is the deltoid region of the upper arm. The patient should be seated with the arm bent at the elbow and the hand on the hip to ensure that the site of administration is prominent.

   Needle gauge and length: A manufacturer-prefilled microinjection syringe is used to administer a 0.1 mL dose into the dermal layer of the skin. The syringe contains a 30-gauge, 1.5-ml microneedle.

4. **Intramuscular route (IM)**: Intramuscular injections are administered into muscle tissue below the dermis and subcutaneous tissue.

   Site: All inactivated vaccines, with the exception of one formulation of meningococcal polysaccharide vaccine (MPSV4), are administered by the intramuscular route. Many inactivated vaccines contain an adjuvant, which is a vaccine component that enhances the immune response to the antigen. Adjuvants can cause an exaggerated local reaction (e.g., pain, swelling, redness) if not injected into the muscle, so proper technique is critical.

   There are only two routinely recommended IM sites for administration of vaccines, the vastus lateralis muscle (anterolateral thigh) and the deltoid muscle (upper arm). Injection at these sites reduces the chance of involving neural or vascular structures. The site depends on the age of the individual and the degree of muscle development. Because there are no large blood vessels in the recommended sites, aspiration before injection of vaccines (i.e., pulling back on the syringe plunger after needle insertion but before injection) is not necessary. A study published in *Archives of Disease in Childhood* in 2007 found that when a vaccine was administered and the needle was withdrawn rapidly without aspiration, there was less evidence of pain than when the vaccine was injected and withdrawn slowly with aspiration. Also, some safety-engineered syringes do not allow for aspiration.

   Needle gauge: 22–25-gauge needle.

   Needle length: The needle should be long enough to reach the muscle mass and prevent the vaccine from seeping into subcutaneous tissue, but not so long as to involve underlying nerves, blood vessels, or bone. The vaccinator should be familiar with the anatomy of the area into which the vaccine will be injected. Decisions on needle size and the site of injection must be made for each person on the basis of the size of the muscle, the thickness of the adipose tissue at the injection site, the volume of the material to be administered, and the injection technique.

5. **Subcutaneous route (subcut)**: Subcutaneous injections are administered into the fatty tissue found below the dermis and above the muscle tissue.

   Site: The recommended subcutaneous sites for vaccine administration are the thigh (for infants younger than 12 months of age) and the upper outer triceps of the arm (for persons 12 months of age and older). If necessary, the upper outer triceps area can be used to administer subcutaneous injections to infants.

   Needle gauge and length: 5/8-in., 23–25-gauge needle.

   Technique.

   - Follow standard medication administration guidelines for site assessment/selection and site preparation.

- To avoid reaching the muscle, pinch up the fatty tissue, insert the needle at a 45-degree angle, and inject the vaccine into the tissue.
- Withdraw the needle and apply light pressure to the injection site for several seconds with a dry cotton ball or gauze.

It is easy to argue that vaccine development represents humankind's most important and successful endeavor [1]. Vaccination has had a tremendous impact on human morbidity and mortality over the years. But routes of vaccine delivery are important and concerning to the development of novel delivery platforms is under way.

References

[1] J. Wallis, D.P. Shenton, R.C. Carlisle, Novel approaches for the design, delivery and administration of vaccine technologies, Clin. Exp. Immunol. 196 (2) (2019) 189–204, https://doi.org/10.1111/cei.13287.

[2] S. Rifo, S. Vella, F. Filiaci, S. Resta, G. Rocchi, Secretory immune response after nasal vaccination with live attenuated influenza viruses, Rhinology 18 (2) (1980) 87–92.

[3] G. Icardi, A. Orsi, A. Ceravolo, F. Ansaldi, Current evidence on intradermal influenza vaccines administered by SoluviaTM licensed micro injection system, Hum. Vaccin. Immunother. 8 (1) (2012) 67–75. G 2012 Landes Bioscience.

Animal models for VL

11

Introduction

Leishman and Donovan first described visceral leishmaniasis (VL) in 1903. This is not to say that leishmaniasis did not exist before 1903. On the contrary, Archibadi in 1922 described an epidemic of Kala-azar, which occurred in the Garo hills of Assam and in Saudi Arabia [1–3]. Also, Cunningham recorded a similar disease that occurred in 1885 caused by a parasite that was later named *Leishmania tropica*; it is the causative agent of cuteneous leishmaniasis. In 1908, Nicolle reported that mammals including dogs could act as reservoir hosts for the leishmania parasite. Leishmania parasite could be transmitted by Phlebotomus sandflies to the human volunteers (www.who.org).

VL is an important health hazard affecting millions of people worldwide. In India, the disease is mainly restricted to Bihar, West Bengal, and northeastern areas of Uttar Pradesh, but now several cases have been reported from Assam and to a lesser extent Tamil Nadu and Orissa. It is a vector-borne infectious disease caused by an intracellular protozoan parasite. It is transmitted to humans by the bite of the female phlebotomy sandfly. It is characterized by fever, often with two daily peaks of hepatosplenomegaly, leukopenia, and cachexia. Extreme emaciation develops, the skin becomes dark, the hair becomes dry, and certain complications such as dysentery, pulmonary tuberculosis, herpes zoster, and epistaxis may occur. The parasite *L. donovani* is characterized by two phases of development, the amastigotes phase in humans and the leptomonad or promastigote phase in the sandfly vector invertebrate host. After an infected blood meal, the vector takes about 4–18 days, according to the species of leishmania, to be able to transmit the infection to the victim. Of the several types of disease, VL is the most dreaded one and is a major cause of mortality and morbidity [4].

A great deal of attention has been focused on VL during the past decade because of the etiological and behavioral factors and the resistance of the parasite to drugs, but largely because of the socioeconomic conditions of the population at risk and the fact that the untreated disease is always fatal. It was estimated that in India, about 1.2 million people were infected with leishmania, resulting in 1000 deaths per year, with most of those belonging to the poor class. The most-serious form of leishmaniasis is VL, which is a life-threatening disease that involves the entire reticuloendothelial system, including the bone marrow and skin where drug delivery becomes difficult. In this disease, the macrophages, which are normally destined to kill any invading or foreign organism, become the site not only for lodging the parasites but also for their multiplication. This results in the depression of the immune status of the host, which drastically alters the management of body as well as drug, filled parasites to further

Visceral Leishmaniasis. https://doi.org/10.1016/B978-0-323-91124-5.00004-7

aggravate the toxic condition leading to derangement of various physiological functions. These conditions favor a suitable situation for intercurrent infections to further worsen the situation. The management of such conditions becomes more complex with existing drug therapy. Because of increasing failure of previously effective drugs with serious toxicities [5, 6]. With the available diagnostic methods, the disease diagnosis will be made at a later stage. Also, monitoring the drug effect is not possible. Hence, this research project has been selected to contribute to the better management of VL by inventing an efficacious, safe, inexpensive, and short-course oral treatment from Ayurvedic medicine for developing and underdeveloped countries.

The development of a potent drug against leishmaniasis is a challenging problem, as the parasites are well protected within the host cells. In addition, due to massive immunosuppression, there is a lack of appropriate host response against the disease [7–9], Consequently, the patient fails to respond satisfactorily to the available antileishmanial drugs. For any compound or agent to be effective, it is necessary that it must reach the parasite in a desired concentration after overcoming the barriers imposed by the host cells. The drug at the same time should have no harmful effect on the host cells and they should boost the immune system [5, 10]. Natural products have good antileishmanial efficacy and are also capable of restoring immune function, which will present an additional advantage in VL diagnosis. For screening of medicinal plants/natural compounds, various in vitro and in vivo techniques have been developed to assess the antileishmanial action [5, 6]. However, a brief overview of animal models and drug testing methods is presented in this chapter. The next chapter explains the various techniques of antileishmanial drug screening in depth that are found to be suitable, reliable, and helpful in drug discovery and development.

Animal models in experimental chemotherapy of VL

It was in 1924 that Smyly and Young succeeded in infecting the Chinese hamster (*Cricetus griseus*). Meleney [11] reported that the histopathology of infected hamsters was similar to that of humans. Alder and Tchernomortes initiated the first chemotherapeutic trial in experimentally induced VL in 1939. Since then, a number of laboratory animals have been tried such as the European hamster (*Citellus purmtairs*) [12], the Macedonian spermophile (*Citellus citellus*) [13], the Syrian or golden hamster (*Mesocricetus auratus*) [14], and cotton rats (*Sigmodon hispidus*). Mikhail and Mansour [15] and McKinney and Hendricks [16] infected the African white-tailed rat (*Mastomys albicaudatus*), which proved to be an excellent host for in vivo maintenance and long-term experimentation with *L. donovani* and *L. braziliensis*. *Mastomys natalelnsis*, a multimammate rat, was first used by Pereira et al. [17] as an experimental model for *L. Mexicana*. Dwivedi et al. [18] successfully used this for the infection of *L. donovani*. Kuwaiti rodents viz. *Musmusculus, Rattus norvegicus, Gerbillus cheesmani, Faculus jaculus* and *Meriones crassus* were used by Taqui and Mohammed [19]. Trotter et al. [20] employed NMRI mice while Neal and Croft [21] used the BALB\c mice.

For secondary screening, much work is under way with various nonhuman primates. In addition to dogs [22], owl monkeys (Aotus trivirgatus) [23], the opposum (Didelphis marsupialis) [24], and the squirrel monkey (Saimiri sciureus) were found susceptible to *Leishmania* infection [23]. Githure *et al.* [25] and Binhazim *et al.* [26] worked on the suitability of a few East African primates for VL such as the vervet monkey (*Cercopithecus aethiops*), the syben monkey (*C. mytis*), and baboons (*Papiocynocephalis*). Recently, langurs (*Presbytis entellus*) have been successfully developed as an acceptable model of *L. donovani* [27].

From the above-cited literature, it appears that for primary screening, hamsters are preferred over NMRI and BALB/c mice. The former is sturdy and can withstand repeated spleen biopsies to check the parasite load, besides having a good intake of *L. donovani* infection [28, 29]. As such, this model has proved suitable for studying the sequential effect of drugs in the same animal. It is desired that in the animal model, the causative organism should interact with the host immune system and also with chemotherapeutic agents comparable to those observed clilnically. The golden hamster fulfills all the required criteria. The experimentation in these animals holds reasonable promise, and if adequate care is exercised, a good correlation could be obtained between experimental results and clinical situations. Regarding secondary/tertiary drug testing for VL, no satisfactory model is available as yet, although langurs bear promise. Various animal models are used in clinical trials. Many animal models are mentioned above, but some animals that are frequently used for VL (Table 11.1) are discussed later more precisely.

A. Mouse

It includes BALB/c, NMRI, DBA/1, C57Bl6, swiss mice, etc.

BALB/c mice are currently used as animal models for drug and vaccine testing. Studies with the mouse model also led to the characterization of immune mechanisms. Another important contribution of the mouse model has been the discovery that chemotherapy is ineffective in the absence of intact T cell-mediated responses. The mouse model has been used widely for the development of vaccines against VL. It is necessary to inject amastigotes intravenously to induce a reproducible pattern of colonization of Leishmania parasite into the liver and spleen.

B. Hamster

The golden hamster was used as an early animal model for the study of VL. Infection with *L. donovani* leads to visceral disease and death, making it a useful tool for the characterization of molecules and mechanisms involved in pathogenesis. It is ideal for drug testing against VL and also for vaccination studies [30,31]. However, because no immunoreagents particularly related to cytokines are available commercially, the studies on this model have been limited. It lacks immunoreagents, making detailed immunological investigation difficult.

Table 11.1 Animal model, VL parasite, and route of administration.

Animal	VL Parasite	Route
Mouse	*L. donovani*	i.v
		i.d.
		i.p.
	L. infantum	i.v
		i.d.
		i.p.
		s.c.
	L. chagasi	i.v.
Hamster	*L. donovani*	i.v
		i.d.
		i.p.
	L. infantum	i.v
		i.d.
		i.p.
		s.c.
	L. chagasi	i.v.
Dog	*L. donovani*	i.v
		i.d.
		i.p.
	L. infantum	i.v
		i.d.
		i.p.
		s.c.
	L. chagasi	i.v.

i.d., intradermal; *i.p.*, intraperitoneal; *i.v.*, intravenous; *s.c.*, subcutaneous.

C. Dog

The dog is the major reservoir of *L. infantum* in the Middle East and Mediterranean regions as well as *L. donovani chagasi* in South America. The disease pattern in dogs and humans is similar. As in humans, the infection remains asymptomatic in some dogs. The dog is a good model for drug and vaccine research against VL.

D. Nonhuman primates

Indian langurs (*Presbytis entellus*) have also been used for the preclinical evaluation of potential antileishmanial drugs and vaccines. *Presbytis entellus* is highly susceptible to single intravenous inoculation of *L. donovani* amastigotes, which invariably produced consistent and progressive acute fatal infection, leading to death between 110 and 150 days postinfection. The infected animals presented all the clinico-immunopathological features as observed in human VL. The availability of a nonhuman primate model would increase the understanding of various aspects of host-parasite interactions.

The success of any research work depends largely on the use of appropriate animal models and reliable techniques. For any kind of experimental work, the availability of these experimental models is desirable, as they provide valuable data on the efficacy of newly developed vaccines and immunotherapeutic agents against leishmaniasis. For immunoprophylaxis, the availability of suitable animal models with physiological, pathological, and immunological responses similar to human VL is of prime importance. When exposed to a variety of *Leishmania* spp. with different pathogenic characteristics, several highly susceptible models for leishmaniasis (fulminating parasite proliferation and clinical VL) have been developed [32] such as in hamsters [33], mice [34], the guinea pig [35], the cotton rat [36], and the dog [37]. Several Old World monkeys such as the vervet monkey and macaques [38] have also been investigated, but this resulted in low and inconsistent infections [39]. Among these, hamsters, mice, and a few nonhuman primates such as the marmoset [26]; the owl monkeys *Aotus irivirgatus* [40]; the squirrel monkey *Saimiri sciureus* [23]; and the Indian langur *Presbytis entellus* [41] have been suggested as potentially useful models for various studies on VL.

With the two-test model, **the** murine model of *L. donovani* infection is a good model of early parasite replication followed by immunological control and subclinical infection, but there is no murine model for the progressive disease observed in active human VL. The hamster developed symptoms and immunologial consequences similar to human VL, resulting in a relentless increase in visceral parasite burden, progressive cachexia, hepatosplenomegaly, pancytopenia, hypergammaglobulinemia, and ultimately death [33]. However, immunological studies in this model are limited by the lack of available immunological reagents. Thus, they hold promise as a primary model for immunotherapeutic/prophylatic trials.

References

[1] P.C. Sen Gupta, A report on kala-azar in Assam, Ind. Med. Gaz. 86 (6) (1951) 266–271.

[2] P.C. Sen Gupta, A report on kala-azar in Assam (concld.), Ind. Med. Gaz. 86 (7) (1951) 312–317.

[3] M.A. al-Zahrani, W. Peters, D.A. Evans, Visceral leishmaniasis in man and dogs in southwest Saudi Arabia, Trans. R. Soc. Trop. Med. Hyg. 82 (6) (1988) 857.

[4] D. Steverding, The history of leishmaniasis, Parasit. Vectors 10 (1) (2017) 82. Published 2017 Feb 15 *https://doi.org/10.1186/s13071-017-2028-5*.

[5] J.W. Mikhail, N.S. Mansour, M.J. Khayyal, Leishmania donovani. : Therapeutic action of antimony dextron glycosides (R.L. : 712) in the golden hamster, Exp. Parasitol. 37 (1975) 348.

[6] L.A. Stauber, E.M. Franchino, J. Grun, An eight day method for screening of compounds against Leishmania donovani in the golden hamster, J. Protozool. 5 (1958) 269.

[7] J. Alexander, D.G. Russell, The Interaction of Leishmania Species With Macrophages, vol. 31, Advances in Parasitology Academic Press, 1992, p. 176.

[8] A. Bertoli, A.V. Greoo, Visceral leishmaniasis presenting with hypertrigycerdaemia, Lancet 11 (1982) 504.

[9] J.K. Nag, M.P. Ghosh, Kala-azar with pulmonary tuberculosis, J. Indian Med. Assoc. 25 (4) (1955) 138.
[10] P.C. Sen Gupta, A. Chatterjee, A.M. Mukherjee, Antimony dextron in the treatment of leishmaniasis, J. Indian Med. Assoc. 37 (12) (1961) 585.
[11] M.E. Meleney, Histopathology of Kala-azar in hamsters, monkey and man, Am. J. Pathol. 1 (1925) 147.
[12] M. Mayer, Empfanglichkelt des curopasi chen hamster (Cricetus frumentarius) for Kala-azar, Arch. Schiffs. V. Tropen. Hyg. 30 (1926) 347.
[13] G. Blanc, J. Comenoptros, Sensibilite du spermophile de Macedoline (Citellus citellus) au Kala-azar Mediterranean, Compt. Revd. Acad. Sci. 191 (1930) 800.
[14] S. Adler, O. Theodor, Investigations on Mediterranean Kala-azar, I-V, Proc. Roy. Soc. Lond. A 108 (1931) 447.
[15] JW Mikhail, NS Mansour, *Mystromys albicaudatus*, the African white-tailed rat, as an experimental host for *Leishmania donovani*, J. Parasitol. 59 (6) (1973) 1085–1087.
[16] L.A. McKinney, L.D. Hendricks, Experimental infection of Mastomys albicaudatus with L. braziliensis, Am. J. Trop. Med. Hyg. 29 (1980) 753.
[17] L.H. Pereira, W. Mayrink, M. Dias, S.M. Michalick, C.A. Costa, P.A. Magalhaes, P. Williams, Mastomys natalensis as an experimental host for L. mexicana, Trans. R. Soc. Trop. Med. Hyg. 72 (1978) 155.
[18] S.N. Dwivedi, P.Y. Guru, R.C. Bhagat, V. Shankhdhar, A.B. Sen, Comparative susceptibility of Mastomys natalensis and golden hamsters (Mesocricetus auratus) to L. donovani (Abstr), in: Proc. V. Natn. Cong. Parasitol., June, 25–27, 1983, p. 99.
[19] M.A. Taqui, A.H.H. Mohammed, Susceptibility of Kuwaity rodents and experimental mice to isolates of Leishmania sp, Trans. R. Soc. Trop. Med. Hyg. 75 (1981) 133.
[20] E.R. Trotter, W. Peters, B.L. Robinson, The experimental chemotherapy of leishmaniasis IV, the development of a rodent model for V.L, Ann. Trop. Med. Parasitol. 74 (1980) 127.
[21] R.A. Neal, S.L. Croft, An in vitro system for determining the activity of compounds against the intracellular amastigote form of L. donovani, J. Antimicrob. Chemother. 14 (1984) 463.
[22] W.L. Chapman Jr., W.L. Hanson, V.B. Waits, K.E. Kinnamon, Antileishmanial activity of selected compounds in dogs experimentally infected with L. donovani, Rev. Inst. Med. Trop. Sao Paulo 21 (1979) 189.
[23] W.L. Chapman Jr., W.L. Hanson, Visceral leishmaniasis in the squirrel monkey (Saimiri sciureus), J. Parasitol. 67 (1981) 40.
[24] W.L. Hanson, W.L. Chapman Jr., L.D. Hendricks, Leishmania donovani in the opossum (Didelphis marsupialis), J. Parasitol. 66 (1980) 700.
[25] J.I. Githure, A.M. Shastry, R. Tarara, J.D. Chulay, M.A. Suleman, C.N. Chunge, J.G. Else, The suitability of East African primates as animal models of visceral leishmaniasis, Trans. R. Soc. Trop. Med. Hyg. 80 (4) (1986) 575.
[26] A.A. Binhazim, S.S. Shin, W.L. Chapman Jr., J.O. Olobo, Comparative susceptibility of African green monkeys (Cercopithecus aethiops) to experimental infection with L. donovani L. d. infantum, Lab. Anim. Sci. 43 (1993) 37.
[27] Anuradha, R. Pal, P.Y. Guru, J.C. Katiyar, N. Sethi, Successful transmission of L. donovani in Indian langurs (Presbytis entellus), Curr. Sci. 59 (2) (1990) 612.
[28] E. Beveridge, F. Hawking, Chemotherapy of leishmaniasis, in: R.J. Schnitzer (Ed.), Experimental Chemotherapy, Academic Press, I, 1963, p. 257.
[29] L.G. Goodwin, The chemotherapy of experimental leishmaniasis II. A dose response curve for the activity of Sodium stibogluconate, Trans R.Soc. Trop. Med. Hyg. 39 (1945) 133.

[30] P. Sharma, S. Rastogi, S. Bhatnagar, J.K. Srivastava, A. Dube, P.Y. Guru, et al., Antileishmanial action of *Tephrosia purpurea* linn, extract and its fractions against experimental visceral leishmaniasis, Drug Dev. Res. 60 (4) (2003) 285–293.

[31] P. Sharma, N. Singh, R. Garg, W. Haq, A. Dube, Efficacy of human beta-casein fragment (54-59) and its synthetic analogue compound 89/215 against *Leishmania donovani* in hamsters, Peptides 25 (11) (2004) 1873–1881.

[32] M. Hommel, C.L. Jaffe, B. Travi, G. Milon, Experimental models for Leishmaniasis and for testing anti leishmanial vaccines, Ann. Trop. Med. Parasitol. 89 (1) (1995) 55.

[33] P.C. Melby, V.V. Tryon, B. Chandrasekar, G.L. Freeman, Cloning of syrian hamster (*Mesocricetus auratus*) cytokine cDNAs and analysis of cytokine mRNA expression in experimental visceral leishmaniasis, Infect. Immun. 66 (1998) 2135–2142.

[34] G.F. Mitchell, Host-protective immunity and its suppression in a parasitic disease: murine cutaneous leishmaniasis, Immunol. Today 5 (1984) 224.

[35] R. Behin, J. Louis, Immune Response to *Leishmania*, Crit. Rev. Trop. Med. 2 (1984) 141.

[36] J.D. Fulton, L.P. Joyner, F.L. Chandler, Studies on Protozoa II. The golden hamster (*Mesocritecetus auratus*) and the cotton rat (*Sigmoden hispidus*) as experimental hosts for *Leishmania donovani*, Trans. R. Soc. Trop. Med. Hyg. 44 (1950) 105.

[37] W. Mayrink, O. Genaro, M. Dias, C.A. Dacasta, M.S.M. Michalick, M.N. Melo, P. Williams, R.T. Dacosta, E. Nascimento, A. Oliveiralima, Vaccination of dogs against *Leishmania (vianna) braziliansis*, Rev. Inst. Med. Trop. Sao Paulo 31 (1989) 67.

[38] V.F. Amaral, V.A.O. Ransatto, F.C. Silva, E. Molinaro, V. Ferreria, S.G. Coutinho, D. - McMahon-Pratt, J.R.G. Grimaldi, Leishmania amazonesis: The Asian Rhesus macaques (*Macaca mulatta*) as an experimental model for study of cutaneous leishmaniasis, Exp. Parasitol. 82 (1996) 34.

[39] T.J. Madindou, Chemotherapy of Visceral leishmaniasis (*Leishmania donovani*) in the squirrel monkey (*Saimiri sciureus*), Ann. Trop. Med. Parasitol. 79 (1) (1985) 13.

[40] W.L. Chapman Jr., W.L. Hanson, L.D. Hendricks, *Leishmania donovani* in the owl monkey (*Aotus trivirgatus*), Trans. R. Soc. Trop. Med. Hyg. 75 (1981) 124.

[41] P.R. Anuradha, K. Zehra, J.C. Katiyar, N. Sethi, G. Bhatia, R.K. Singh, The Indian langur: preliminary report of a new non-human primate host for visceral leishmaniasis, Bull. World Health Organ. 70 (1992) 63.

Vaccination approaches for VL 12

Various vaccination approaches explored in VL

We have discussed the common vaccination approach in general in Chapters 7 and 9. Among those approaches, several tools/strategies for the design and delivery of vaccines used for visceral leishmaniasis (VL) (Table 12.1) are described in this section. The development of VL vaccines may be classified into:

A. First-generation vaccines

This includes trials using the whole living virulent VL organism, the attenuated or killed promastigote form of VL, and they were used with or without adjuvants. The early trials with killed Leishmania as a vaccine were conducted in Brazil in the 1940s. Later, from the 1970s onward, a research group developed a killed vaccine composed of five isolates of Leishmania containing four different species [48].

B. Second-generation vaccines

This includes trials using different recombinant molecules, either parasite fractions or genetically engineered organisms, that is, *Leishmania* made avirulant by removing specific genes; bacteria or viruses carrying and expressing leishmanial genes; or even synthetic peptides representing epitopes of antigens. There has been much activity in the discovery of candidates for second-generation vaccines in leishmaniasis vaccine research [49].

C. Third-generation vaccines

A DNA vaccine is called a third-generation vaccine. It contains DNA that codes for specific proteins (antigens) from a pathogen. The DNA is injected into the body and taken up by cells, whose normal metabolic processes synthesize proteins based on the genetic code in the plasmid that they have taken up. In documenting the beneficial role of CD8+ T cells in VL treatment and prevention, some research has focused on DNA vaccines. In a very recent study, it was shown that a third-generation vaccine employing simian adenovirus (ChAd63) could effectively elicit a wide range of CD8+ T cells, specified for Leishmania antigens [50].

A. First-generation vaccines

The whole parasitic agent is given as a vaccine, either in the killed mode or the attenuated mode. They are of the following types:

1. Inactivated or killed vaccines

A cocktail of killed *Leishmania* stocks or a single strain induces significant production. It develops long-lasting specific T-cell responses of the Th1 type in the host with the production of IFN-γ and the absence of IL-4. DTH skin test conversion can be used

Visceral Leishmaniasis. https://doi.org/10.1016/B978-0-323-91124-5.00002-3

Table 12.1 Vaccination approaches that have been used for VL vaccine development.

1. Killed vaccines		
Vaccine candidate	**Evaluated in parasite-host system**	**References**
Killed promastigote *L. donovani* + glucan	*L. donovani*/Mice	[1]
Autoclaved *L. donovani* + BCG	*L. donovani*/Hamster	[2]
10^8 formaldehyde killed promastigotes of *L. donovani*	*L. donovani*/Hamster/Mice	[3]
Autoclaved *L. major* + BCG	*L. donovani*/Hamster/ Langur/Human	[4] [5] [6] [7]
	L. infantum/Dog	
Killed promastigote of *L.infantum* + FCA	*L. infantum*/Dog	[8]
Merthiolated promastigote of *L. brazilensis* + BCG	*L. chagasi*/Dog	[9]
BCG alone	*L. donovani*/hamster	[10]
2. Live attenuated/nonattenuated vaccines		
Vaccine candidate	**Evaluated in parasite-host system**	**References**
Live promastigote of *L. tropica*	*L. donovani*/Human	[11]
Live promastigote of *L. donovani*	*L. donovani*/Human	[12]
	L. donovani/Hamster	[3]
10^7 live amastigotes of *L. donovani*	*L. donovani*/Hamster	[13]
10^7 live promastigotes of *L. donovani*	*L. chagasi*/Mice	[14]
Centrin protein lacking its N-terminal domain of *L. donovani*	*L. donovani*/Human	[15]
3. Recombinant protein vaccines		
Vaccine/strain	**Evaluated in parasite-host system**	**References**
FML	*L. donovani*/Mice/ Hamster	[16] [17]
LiESAp	*L. infantum*/Dog	[18] [19]
A2	*L. donovani*/Mice	[20]
rHASPB	*L. donovani*/Mice	[21]
dp72 + *C. parvum*	*L. donovani*/Mice	[22]
FML+ FIA/BCG/saponin/IL-12/QS 21/Quil A/Aluminum hydroxide	*L. infantum*/Mice/ Hamster/Dog	[23–26]
Gp36	*L. donovani*/Mice	[27]
Eluted gp63 in positive liposome	*L. donovani*/Mice	[28]
Q protein formulated with BCG	*L. infantum*/Dog	[29]

Table 12.1 Continued

3. Recombinant protein vaccines		
Vaccine/strain	**Evaluated in parasite-host system**	**References**
LiF2 (*L.infantum* fraction)	*L. infantum*/Dog	[30]
Glucose regulated protein-78	*L. donovani*/Mice	[31]
gp-63 protein	*L. donovani*/Mice	[32]
Integral membrane protein + CFA	*L. donovani*/Hamster	[33]
BCG expressing flagellar antigen LCR1/	*L. chagasi*/Mice	[34]
Recombinant LCR1		[14]
5×10^7 promastigote BT1 Null mutant	*L. donovani*/Mice	[35]
IL-12 p40 –p35 fusion cDNA	*L. donovani*/Mice	[36]
4. DNA vaccines		
Vaccine/strain	**Evaluated in parasite-host system**	**References**
ORFF	*L. donovani*/Mice	[37]
A2	*L. donovani*/Mice	[38]
A2 and NH	*L. donovani*/Mice	[39]
KMP11	*L. donovani*/Hamster	[40]
PPG	*L. donovani*/Hamster	[41]
rHASPB1+ IL12	*L. donovani*/Mice	[42]
Gammaglutamylcysteine synthetase	*L. donovani*/Mice	[43]
p36 (LACK) DNA	*L. chagasi*/Mice	[44]
p36 (LACK) DNA	*L. donovani*/Mice	[36]
cDNA sublibraries	*L. donovani*/Mice	[36]
Only papLe 22 cDNA	*L. donovani*/Hamster	[45]
5. Nonprotein antigen		
Vaccine/strain	**Evaluated in parasite-host system**	**References**
Some fraction of *L. donovani* + β-1,3-glucan	*L. donovani*/Mice	[46]
Purified glycolipid + BCG	*L. donovani*/Hamster and mouse	[47]

as a surrogate marker to check it. The use of living vaccines, or leishmanization, is based on convalescent immunity, which is acquired following the induction of a lesion at a selected site with a small dose of a cutaneous strain. Leishmanization has been practiced for hundreds of years, especially in the Middle East and parts of the Soviet Union. It is a currently used practice in Iran, where a massive program was begun in 1982 after outbreaks of cutaneous leishmaniasis (CL) among military personnel. As shown in the Soviet Union and Iran, leishmanization can protect 90% or more of the population from multiple cutaneous lesions or lesions on the face [51]. Unfortunately, leishmanization using virulent strains can be associated with several problems, including the development of large slowly healing lesions, lesion contamination, and exacerbation of other dermatologic conditions. For these reasons, the practice has been restricted in the Soviet Union and discontinued in Israel and other countries.

2. Live-attenuated or weakened vaccine

Avirulent organisms, irradiated parasites, and genes encoding the enzyme responsible for pathogenesis are used in such cases. It induces the closest mimic to the natural course of infection and therefore may lead to similar immune responses. Because of the small load of antigen delivered by the transient infection, immune responses may be skewed even more toward a Th1 protective response than in a natural infection. Such vaccines are further subcategorized into:

a. *Without adjuvants*

Initial experiments used sonicated promastigotes as crude antigens for vaccines, as done by Handman et al. [52]. Although these induced some protection against *L. major* in genetically resistant strains of mice, they had little effect on the susceptible strains. Moreover, the immunized individuals developed long-lasting specific T-cell responses of the Th 1 type, which may indicate a potential to protect from infection.

In Iran, Ali and Afrin [53] prepared a vaccine against *Leishmania donovani*. They used a leishmanial antigen to immunize BALB/C mice by the intraperitoneal injection of 20 micrograms of the antigen in phosphate-buffered saline (PBS). They received a significant level of protection against the disease for up to 4 months. An interesting observation is that this vaccine produces a predominantly CD8+ T-cell response like that seen in leishmaniasis patients after successful treatment.

b. *With adjuvants*

In India in 1998, Dube et al. attempted the vaccination of langur monkeys against L. *donovani* by using autoclaved *L. major* plus bacilllus Calmette-Guerin (BCG). Triple doses each of 1 mg of killed *L. major* (ALM) plus 1 mg of BCG intradermally were tested and found effective for approximately 8 months. This study indicated that a combination of ALM plus BCG may be a good candidate vaccine against human VL. In Iran, a mixed BCG *L. major* killed vaccine has undergone clinical trials for safety and efficacy. In one study, there was little difference in disease incidence between the group vaccinated with BCG alone and the group given BCG and the vaccine. A second study showed that in the longer term, the vaccine combination provided better efficacy than BCG alone, suggesting that BCG may have had only a transient immunostimulatory effect [54,55]. Vaccination with a single dose of 1 mg of *L. major* protein and BCG is the simplest vaccine tested so far. Although it proved safe, only about 35% of vaccinated individuals became skin test positive. If skin test conversion is a surrogate marker for protection, the efficacy of this vaccine is not remarkable. It may require multiple doses to increase immunogenicity [56]. However, because delayed-type hypersensitivity was not predictive of protection in this monkey model, the value of delayed-type hypersensitivity as a surrogate marker of protection in humans needs to be reassessed [56].

B. Second-generation vaccines

Delivery of many more parasite antigen is possible with subunit or recombinant antigens. The second-generation VL vaccines are still in the preclinical/clinical stage [57]. They are grouped into:

1. Purified protein antigens or synthetic peptides/proteins

T-cells recognize peptides derived from cytosolic proteins bound in the MHC class I groove and peptides derived from the lysosomal compartment bound in the MHC class II groove on the antigen presenting cell (APC) surface. It appears that virtually any parasite protein might function as an antigen, regardless of its location in the parasite. Recombinant antigens can be delivered as purified proteins, as the naked DNA encoding them is used. Before that, the identification of interesting and effective antigens is required. The most interesting approach for the identification of potential vaccine candidates has been the elution of antigenic peptides from APC. Several Leishmania gp63 peptides were tested successfully in animal models. Host protection was long-lasting in the experiment, indicating the induction of long-term T-cell memory.

The 1980 century was concerned for the use of peptide vaccines [58]. This seems to have waned in recent times, and the focus appears to have moved to the use of recombinant DNA-produced polypeptides and to naked DNA. Some considerations make peptide antigens less attractive: (1) the magnitude of the T-cell memory induced, (2) the inability of all individuals in the populations to respond to the peptide, and (3) the economics of its production. Because the antigenic peptide is processed and presented to T cells in the context of MHC class I or class II but all peptides are not associate with all MHC types and few peptides are not recognized by all individuals in the population. There are additional "holes" in the ability to respond to individual peptides due to the failure of processing or transport or the deletion of certain T-cell specificities (due to self-tolerance). Host protection should be long-lasting, further indicating the induction of long-term T-cell memory. Sjölander et al. [59] recorded the success of subunit vaccines in plasmodium based on recombinant proteins or peptides. Several factors may account for this. Some polypeptides such as PSA-2 need to be in their native conformation for antigen processing, and *E. coli*-derived recombinant proteins may not fulfill this requirement. This problem may be overcome by the exploitation of the parasites themselves by the overexpression of parasite antigens in transfected nonpathogenic Leishmania strains or in the trypanosomatid [60]. Apparently, polypeptides expressed in these systems will be abundant, correctly folded, and glycosylated. Another reason for the low success rate of subunit vaccines is that some polypeptides may be minor immunogens. So, even though they may be excellent in a cocktail vaccine, individually they may provide only partial protection. The immune responses in leishmaniasis can range from protective to positively harmful. These differences in the quality of the response are at least partly due to the predominance of Th1 or Th2 cytokines, and may be greatly influenced by the antigen dose. Accordingly, the amount of antigen and possibly the route of its administration may be important issues [56].

Another thorny issue concerns adjuvants. The delivery system may be critical in biasing the type of T-cell response induced, and this can determine whether protection is achieved or whether immunization makes the disease worse. The delivery of PSA-2 packaged in immunostimulant complexes induced a strong but mixed Th1/Th2 response and no protection. However, its delivery as a DNA vaccine induced a low but exclusive Th1 response and protection. In most experimental systems, adjuvants are essential to provoke protective immunity. However, the most effective adjuvants generally cause strong inflammation, which may be essential for adjuvanticity but may preclude their use in humans because of unacceptable side effects [56].

2. Nonprotein antigens

Early studies on vaccine development indicated that glycolipids such as the *Leishmania* lipophosphoglycan (LPG) provided excellent protection. Protection depended on the use of adjuvants such as liposomes or *Corynebacterium parvum* and on the integrity of the molecules. Not only was the water-soluble form of LPG lacking the glycosylphosphatidylinostiol anchor not protective, but it exacerbated the disease [61]. At the time when that work was published, the immune mechanism leading to host protection by such a nonprotein molecule was totally mysterious. Immunity was known to be T-cell-mediated, but T cells were not thought to recognize or present nonprotein antigens. Today, it is accepted that many novel and interesting microbial antigens, including mycobacterial glycolipids, can be recognized by T cells and that these antigens are presented to T cells by a special subset of MHC class I proteins known as CD1. In this context, it may be rewarding to reevaluate the potential of LPG as a vaccine candidate [56].

Recombinant proteins or peptides have poor or variable efficacy. Some polypeptides such as PSA-2 need to be in their native conformation for antigen processing; *E. coli*-derived recombinant proteins may not fulfill this requirement. This problem may be overcome by the exploitation of parasites themselves by the overexpression of parasite antigen in transfected nonpathogenic Leishmania strains. Probably, polypeptides expressed in these systems will be abundant, correctly folded, and glycosylated. Some polypeptides may be minor immunogens and so even though they may be excellent in a cocktail vaccine, individually they may provide only partial protection. The delivery systems (adjuvants) may be critical in biasing the type of T-cell responses. The delivery of proteins/peptides packaged in immunostimulant complexes induced a strong but mixed Th1/Th2 response and no protection, whereas delivery as a DNA vaccine induced a low but exclusive Th1 response and protection.

C. Third-generation vaccines

Third-generation vaccines are advantageous over first- and second-generation vaccines. Their types are described below.

1. Recombinant DNA vaccines

These vaccines comprise recombinant DNA-derived target antigens. Target antigens are species-specific and life cycle stage-specific. Leishmania can invade and undergo a limited number of replications in macrophages without producing disease. Recent advances in the ability to manipulate the genome of *Leishmania* by eliminating or introducing genes could make it possible to produce a stable mutant that would at least cause a futile infection if inoculated into the host. Such infections should produce a protective immune response in humans like that seen in a self-healing natural infection as well as life-long immunity [62]. This can be done by removing the essential genes of Leishmania for long-term survival in the mammalian host (e.g., dihydrofolate reductase-thymidylate synthetase (DHFR-TS), LPG, PPG, cysteine proteases cpa and cpb, etc.). These candidates could provide protection to virulent *Leishmania*. It has been reported that these attenuated parasites may be used to induce long-term protection against leishmaniasis, and that the immunity induced can be maintained in the absence of a strong Th1 response.

The use of attenuated organisms is very attractive because they are the closest mimic to the natural course of infection and may therefore lead to similar immune responses. Moreover, because of the small load of antigen delivered by the transient infection, the immune responses may be skewed even more toward a Th1 protective response than in natural infection [35]. Such immunization will also deliver many more parasite antigens than the limited number possible with subunit or recombinant antigens. Summarizing a large amount of experimental evidence, Rivier et al. [63] concluded that the injection of attenuated organisms achieved better protection than any method involving recombinant gp^{63} as test antigen delivered with a variety of adjuvants and delivery systems. If this conclusion is shown to be generally applicable to other vaccine candidates, the prospect of using attenuated *Leishmania* vaccines in preference to subunit or recombinant approaches will gain favor. The disadvantages of such vaccines are the logistics of their large-scale production and distribution in the field.

2. Naked DNA vaccines

Immunization with naked DNA is a new approach that promises to revolutionize the prevention and treatment of infectious diseases [64]. The gene encoding the vaccine candidate is cloned in a mammalian expression vector, and the DNA is injected directly into the muscle or skin. Surprisingly, the plasmid DNA is taken up by cells and translocated to the nucleus, where it is transcribed into RNA and then translated in the cytoplasm. The efficiency of the uptake and expression of plasmid DNA must be extremely low, but it is enough to provoke immune responses in both T and B cells [64]. Antigen presentation is not done directly via skin or muscle cells, but is done via either class I or class II MHC molecules on professional antigen-presenting cells [65]. Responses mediated by both -$CD4^+$ and -$CD8^+$ are induced, making a DNA vaccine attractive for a *Leishmania* vaccine. In addition to being able to induce the appropriate immune responses, DNA vaccines are attractive because they ensure the appropriate folding of the polypeptide, produce the antigen over long periods, and do not require adjuvants. Another advantage is that the technology for production is very simple. DNA is stable, has a long shelf life, and does not require a strict cold chain for distribution. Concerns raised in relation to safety, such as the integration of the DNA into the mammalian genome and the induction of autoimmune disease or cancer, have not been substantiated to date [56]. Vaccinations with DNA encoding gp63, LACK, and PSA-2 all protected both genetically resistant and susceptible mice from infection with *L. major* [59,64]. Protection was accompanied by Th1 immune responses. Unexpectedly, protection induced by LACK depended on $CD8^-$ T cells, and depletion of this population abrogated protection. A comparative evaluation of different DNA vaccine candidates against experimental murine leishmaniasis due to *L. major* was done by Ahmed et al. [66]. The candidate DNA vaccines encode to the following antigens: LACK, PSA2, LeIF, and two newly identified P20 and ribosomal-like proteins. All were found to be protective, but the most promising gene was LACK.

Drugs or vaccines for control of VL

Significant advances have been made in the development of anti-*Leishmania* drugs. In addition to the traditional pentavalent antimonial, new drugs such as aminosidine and liposome-delivered amphotericin B have been introduced. These drugs are currently used in chemotherapy. Other drugs such as allopurinol and ketoconazole are in advanced stages of research that target parasite-specific biosynthetic pathways. There are advantages to the use of chemotherapy for the control of leishmaniasis. Drugs are not affected by parasite heterogeneity. They can be administered noninvasively, that is, orally. The most important thing is that once a drug is developed, better formulations are relatively easy to produce to enhance their efficacy. But unfortunately, drugs are less effective in immunocompromised persons and drug resistance is another impending major problem.

Unlike drugs or chemotherapy, vaccination is usually a "one-shot" event. This makes it cheaper, and the easier logistics of administration lead to much better compliance. Vaccines can be administered in both prophylactic and therapeutic modes. Immunotherapy has proved successful in both human and experimental VL. An additional advantage of vaccination is its long-lasting effect. It also avoids problems such as drug resistance.

Advantages of drugs

1. Drugs are not affected by parasite heterogeneity.
2. Drugs are administered orally.
3. Once the drugs are developed, better formulations are relatively easy to produce.

Disadvantages of drugs

1. Drugs are much less effective in immunocompromised individuals.
2. Drug resistance can occur.

Advantages of vaccines

1. Usually a "one-shot" affair.
2. Cheaper and easier logistics of administration lead to much better compliance.
3. They can be administered in both prophylactic and therapeutic models.

4. Long-lasting effect, and they avoid problems with drug resistance.
5. Prevents epidemics and pandemics.
6. Prevents people spreading the disease to others.

Disadvantages of vaccines

1. Not guaranteed to work or provide 100% protection.
2. Possible side effects.
3. Invasive and possibly painful.
4. Booster injections are inconvenient.

A combination of chemotherapy, immunotherapy, and immunoprophylaxis could be ideal in the battle against VL.

References

[1] T.W. Holbrook, J.A. Cook, B.W. Parker, Immunization against *Leishmania donovani*: glucan as an adjuvant with killed promastigotes, Am. J. Trop. Med. Hyg. 30 (1981) 762–768.

[2] J.K. Srivastava, A. Misra, P. Sharma, B. Srivastava, S. Naik, A. Dube, Prophylactic potential of autoclaved *Leishmania donovani* with BCG against experimental visceral leishmaniasis, Parasitology 127 (Pt 2) (2003) 107–114.

[3] S. Mukhopadhyay, P. Sen, S. Bhattacharya, S. Mazumdar, S. Roy, Immunoprophylaxis and immunotherapy against experimental visceral leishmaniasis, Vaccine 17 (1998) 289–296.

[4] A. Dube, P. Sharma, J.K. Srivastava, A. Misra, S. Naik, J.C. Katiyar, Vaccination of langur monkeys (*Presbytis entellus*) against *Leishmania donovani* with autoclaved *L. major* plus BCG, Parasitology 116 (Pt 3) (1998) 219–221.

[5] A. Misra, A. Dube, B. Srivastava, P. Sharma, J.K. Srivastava, J.C. Katiyar, S. Naik, Successful vaccination against *Leishmania donovani* infection in Indian langur using alum-precipitated autoclaved *Leishmania major* with BCG, Vaccine 19 (2001) 3485–3492.

[6] E.A. Khalil, A.M. El Hassan, E.E. Zijlstra, M.M. Mukhtar, H.W. Ghalib, B. Musa, M.E. Ibrahim, A.A. Kamil, M. Elsheikh, A. Babiker, F. Modabber, Autoclaved *Leishmania major* vaccine for prevention of visceral leishmaniasis: a randomised, double-blind, BCG-controlled trial in Sudan, Lancet 356 (2000) 1565–1569.

[7] A. Rhalem, H. Sahibi, N. Guessous-Idrissi, S. Lasri, A. Natami, et al., Immune response against Leishmania antigens in dogs naturally and experimentally infected with *Leishmania infantum*, Vet. Parasitol. 81 (1999) 173–184.

[8] J.H.L. Ferreira, L.S. Silva, I.M. Longo-Maugéri, S. Katz, C.L. Barbiéri, Use of a recombinant cysteine proteinase from *Leishmania (Leishmania) infantum* chagasi for the immunotherapy of canine visceral Leishmaniasis, PLoS Negl. Trop. Dis. 8 (3) (2014), e2729.

[9] W. Mayrink, O. Genaro, J.C. Silva, R.T. da Costa, W.L. Tafuri, V.P. Toledo, et al., Phase I and II open clinical trials of a vaccine against *Leishmania chagasi* infections in dogs, Mem. Inst. Oswaldo Cruz 91 (1996) 695–697.

[10] J.C. Jarecki-Black, A.B. Glassman, T.W. Holbrook, The effect of BCG-vaccine upon experimental visceral leishmaniasis in hamsters, Ann. Clin. Lab. Sci. 14 (6) (1984) 464–466.

[11] D.J. Wyler, F.I. Weinbaum, H.R. Herrod, Characterization of in vitro proliferative responses of human lymphocytes to leishmanial antigens, J. Infect. Dis. 140 (1979) 215–221.
[12] P.E. Manson-Bahr, R.B. Heisch, Transient infection of man with a Leishmania (*L. adleri*) of lizards, Ann. Trop. Med. Parasitol. 55 (1961) 381–382.
[13] T.J. Nolan, J.P. Farell, Experimental infections of the multi-mammate rat (*Mastomys natalensis*) with *Leishmania donovani* and *Leishmania major*, Am. J. Trop. Med. Hyg. 36 (1987) 264–269.
[14] J.A. Streit, T.J. Recker, F.G. Filho, S.M. Beverley, M.E. Wilson, Protective immunity against the protozoan *Leishmania chagasi* is induced by subclinical cutaneous infection with virulent but not avirulent organisms, J. Immunol. 166 (2001) 1921–1929.
[15] A. Selvapandiyan, A. Debrabant, R. Duncan, J. Muller, P. Salotra, G. Sreenivas, J.L. Salisbury, H.L. Nakhasi, Centrin gene disruption impairs stage-specific basal body duplication and cell cycle progression in Leishmania, J. Biol. Chem. 279 (2004) 25703–25710.
[16] E.M. Saraiva, B.A. de Figueiredo, F.N. Santos, G.P. Borja-Cabrera, D. Nico, L.O. Souza, M.-A.C. de Oliveira, E.P. de Souza, P. Fampa, L.E. Parra, I. Menz, J.G. Dias, S.M. de Oliveira, C.B. Palatnik-de-Sousa, The FML-vaccine (Leishmune) against canine visceral leishmaniasis: a transmission blocking vaccine, Vaccine 24 (2006) 2423–2431.
[17] V.O. Da Silva, G.P. Borja-Cabrera, N.N.C. Pontes, et al., A phase III trial of efficacy of the FML-vaccine against canine kala-azar in an endemic area of Brazil (São Gonçalo do Amaranto, RN), Vaccine 19 (2001) 1082–1092.
[18] G. Bourdoiseau, C. Hugnet, R.B. Goncalves, et al., Effective humoral and cellular immunoprotective responses in LiESAp-MDP vaccinated protected dogs, Vet. Immunol. Immunopathol. 128 (2009) 71–78.
[19] J.L. Lemesre, P. Holzmuller, R.B. Goncalves, et al., Long-lasting protection against canine visceral leishmaniasis using the LiESAp-MDP vaccine in endemic areas of France: doubleblind randomised efficacy field trial, Vaccine 25 (2007) 4223–4234.
[20] W.W. Zhang, G. Matlashewski, Loss of virulence in *Leishmania donovani* deficient in an amastigote-specific protein, A2, Proc. Natl. Acad. Sci. U. S. A. 94 (16) (1997) 8807–8811, https://doi.org/10.1073/pnas.94.16.8807.
[21] E. McFarlane, T. Mokgethi, P.M. Kaye, et al., IL-4 mediated resistance of BALB/c mice to visceral Leishmaniasis is independent of IL-4Rα signaling via T cells, Front. Immunol. 10 (2019) 1957.
[22] N. Rachamim, C.L. Jaffe, Pure protein from *Leishmania donovani* protects mice against both cutaneous and visceral leishmaniasis, J. Immunol. 150 (1993) 2322–2331.
[23] J.M. Da Costa, A.B. Neogy, I. Vouldoukis, M.L. Silva, M. Gentilini, L. Monjour, Antigenic components of partially purified antigens of *Leishmania donovani* infantum recognized by sera from dogs with asymptomatic or active visceral leishmaniasis, Am. J. Trop. Med. Hyg. 55 (1996) 511–515.
[24] W.R. Santos, R.R. Bernardo, L.M. Pecanha, M. Palatnik, J.P. Parente, C.B. Palatnik de Sousa, Haemolytic activities of plant saponins and adjuvants. Effect of *Periandra mediterranea* saponin on the humoral response to the FML antigen of *Leishmania donovani*, Vaccine 15 (1997) 1024–1029.
[25] W.R. Santos, V.M. de Lima, E.P. de Souza, R.R. Bernardo, M. Palatnik, C.B. Palatnik de Sousa, Saponins, IL12 and BCG adjuvant in the FML-vaccine formulation against murine visceral leishmaniasis, Vaccine 21 (2002) 30–43.
[26] W.R. Santos, I.A. Aguiar, E. Paraguai de Souza, V.M. de Lima, M. Palatnik, C.B. Palatnik-de-Sousa, Immunotherapy against murine experimental visceral leishmaniasis with the FML-vaccine, Vaccine 21 (2003) 4668–4676.

[27] E. Paraguai de Souza, R.R. Bernardo, M. Palatnik, C.B. Palatnik de Sousa, Vaccination of Balb/c mice against experimental visceral leishmaniasis with the GP36 glycoprotein antigen of *Leishmania donovani*, Vaccine 19 (23–24) (2001) 3104–3115.
[28] F. Afrin, R. Rajesh, K. Anam, M. Gopinath, S. Pal, N. Ali, Characterization of *Leishmania donovani* antigens encapsulated in liposomes that induce protective immunity in BALB/c mice, Infect. Immun. 70 (2002) 6697–6706.
[29] L.C. Afonso, T.M. Scharton, L.Q. Vieira, M. Wysocka, G. Trinchieri, P. Scott, The adjuvant effect of interleukin-12 in a vaccine against *Leishmania major*, Science 263 (1994) 235–237.
[30] S. Dunan, D. Frommel, L. Monjour, B.W. Ogunkolade, A. Cruz, M. Quilici, Vaccination trial against canine visceral leishmaniasis, Parasite Immunol. 11 (1989) 397–402.
[31] D.G. Russell, The macrophage-attachment glycoprotein gp63 is the predominant C3-acceptor site on *Leishmania mexicana* promastigotes, Eur. J. Biochem. 164 (1987) 213–221.
[32] D.G. Russell, J. Alexander, Effective immunization against cutaneous leishmaniasis with defined membrane antigens reconstituted into liposomes, J. Immunol. 140 (1988) 1274–1279.
[33] R. Garg, J.K. Srivastava, A. Pal, S. Naik, A. Dube, Isolation of integral membrane proteins of *Leishmania promastigotes* and evaluation of their prophylactic potential in hamsters against experimental visceral leishmaniasis, Vaccine 23 (2005) 1189–1196.
[34] J.A. Streit, T.J. Recker, J.E. Donelson, M.E. Wilson, BCG expressing LCR1 of *Leishmania chagasi* induces protective immunity in susceptible mice, Exp. Parasitol. 94 (2000) 33–41.
[35] B. Papadopoulou, G. Roy, M. Breton, et al., Reduced infectivity of a Leishmania donovani biopterin transporter genetic mutant and its use as an attenuated strain for vaccination, Infect. Immun. 70 (1) (2002) 62–68.
[36] P.C. Melby, J. Yang, W. Zhao, L.E. Perez, J. Cheng, *Leishmania donovani* p36 (LACK) DNA vaccine is highly immunogenic but not protective against experimental visceral leishmaniasis, Infect. Immun. 69 (2001) 4719–4725.
[37] B. Sukumaran, P. Tewary, S. Saxena, R. Madhubala, Vaccination with DNA encoding ORFF antigen confers protective immunity in mice infected with *Leishmania donovani*, Vaccine 21 (2003) 1292–1299.
[38] A.P. Fernandes, M.M.S. Costa, E.A.F. Coelho, et al., Protective immunity against challenge with *Leishmania (Leishmania) chagasi* in beagle dogs vaccinated with recombinant A2 protein, Vaccine 26 (46) (2008) 5888–5895.
[39] F.H.C. Zanin, E.A.F. Coelho, C.A.P. Tavares, et al., Evaluation of immune responses and protection induced by A2 and nucleoside hydrolase (NH) DNA vaccines against *Leishmania chagasi* and *Leishmania amazonensis* experimental infections, Microbes Infect. 9 (9) (2007) 1070–1077.
[40] R. Basu, S. Bhaumik, J.M. Basu, K. Naskar, T. De, S. Roy, Kinetoplastid membrane protein-11 DNA vaccination induces complete protection against both pentavalent antimonial-sensitive and -resistant strains of *Leishmania donovani* that correlates with inducible nitric oxide synthase activity and IL-4 generation: evidence for mixed Th1- and Th2-like responses in visceral leishmaniasis, J. Immunol. 174 (11) (2005) 7160–7171.
[41] M. Samant, R. Gupta, S. Kumari, P. Misra, P. Khare, P.K. Kushawaha, A.A. Sahasrabuddhe, A. Dube, Immunization with the DNA-encoding N-terminal domain of proteophosphoglycan of *Leishmania donovani* generates Th1-type immunoprotective response against experimental visceral leishmaniasis, J. Immunol. 183 (1) (2009) 470–479.

[42] S. Stäger, D.F. Smith, P.M. Kaye, Immunization with a recombinant stage-regulated surface protein from *Leishmania donovani* induces protection against visceral leishmaniasis, J. Immunol. 165 (12) (2000) 7064–7071.

[43] A. Mukherjee, P.K. Padmanabhan, S. Singh, G. Roy, I. Girard, M. Chatterjee, M. Ouellette, R. Madhubala, Role of ABC transporter MRPA, gamma-glutamylcysteine synthetase and ornithine decarboxylase in natural antimony-resistant isolates of *Leishmania donovani*, J. Antimicrob. Chemother. 59 (2) (2007) 204–211.

[44] E.A. Marques-da-Silva, E.A. Coelho, D.C. Gomes, M.C. Vilela, C.Z. Masioli, C.A. Tavares, et al., Intramuscular immunization with p36(LACK) DNA vaccine induces IFNgamma production but does not protect BALB/c mice against *Leishmania chagasi* intravenous challenge, Parasitol. Res. 98 (2005) 67–74.

[45] I. Suffia, B. Ferrua, X. Stien, B. Mograbi, P. Marty, D. Rousseau, K. Fragaki, J. Kubar, A novel *Leishmania infantum* recombinant antigen which elicits interleukin 10 production by peripheral blood mononuclear cells of patients with visceral leishmaniasis, Infect. Immun. 68 (2) (2000) 630–636.

[46] J.A. Cook, T.W. Holbrook, Immunogenicity of soluble and particulate antigens from *Leishmania donovani* effect of glucan as an adjuvant, Infect. Immun. 40 (3) (1983) 1038–1043.

[47] W.K. Tonui, S.S. Mpoke, A.S. Orago, S.J. Turco, P.A. Mbati, G.M. Mkoji, Leishmania donovani-derived lipophosphoglycan plus BCG induces a Th1 type immune response but does not protect Syrian golden hamsters (*Mesocricetus auratus*) and BALB/c mice against *Leishmania donovani*, Onderstepoort J. Vet. Res. 70 (4) (2003) 255–263.

[48] O. Genaro, V.P. de Toledo, C.A. Da Costa, M.V. Hermeto, L.C. Afonso, W. Mayrink, Vaccine for prophylaxis and immunotherapy, Brazil. Clin. Dermatol. 14 (1996) 503–512.

[49] N.H. Lee, J.A. Lee, S.Y. Park, C.S. Song, I.S. Choi, J.B. Lee, A review of vaccine development and research for industry animals in Korea, Clin. Exp. Vaccine Res. 1 (1) (2012) 18–34.

[50] M. Osman, A. Mistry, A. Keding, R. Gabe, E. Cook, S. Forrester, et al., A third generation vaccine for human visceral leishmaniasis and post kala azar dermal leishmaniasis: first-in-human trial of ChAd63-KH, PLoS Negl. Trop. Dis. 11 (2017), e0005527.

[51] A. Khamesipour, Y. Dowlati, A. Asilian, R. Hashemi-Fesharki, A. Javadi, S. Noazin, F. Modabber, Leishmanization: use of an old method for evaluation of candidate vaccines against leishmaniasis, Vaccine 23 (28) (2005) 3642–3648.

[52] E. Handman, R. Ceredig, G.F. Mitchell, Murine cutaneous leishmaniasis: disease patterns in intact and nude mice of various genotypes and examination of some difference between normal and infected macrophages, Aust. J. Exp. Biol. Med. Sci. 57 (1979) 9–29.

[53] N. Ali, F. Afrin, Protection of mice against visceral leishmaniasis by immunization with promastigote antigen incorporated in liposomes, J. Parasitol. 83 (1997) 70–75.

[54] I. Sharifi, A.R. Fekri, M.R. Aflatoonian, et al., Randomized vaccine trial of single dose of killed *Leishmania major* plus BCG against anthroponotic cutaneous leishmaniasis in Bam, Iran, Lancet 351 (1998) 1540–1544.

[55] A.Z. Momeni, T. Jalayer, M. Emamjomeh, A. Khamesipour, F. Zicker, R.L. Ghassemi, Y. Dowlati, I. Sharifi, M. Aminjavaheri, A. Shafiei, M.H. Alimohammadian, R. Hashemi-Fesharki, K. Nasseri, T. Godal, P.G. Smith, F. Modabber, A randomised, double-blind, controlled trial of a killed *L. major* vaccine plus BCG against zoonotic cutaneous leishmaniasis in Iran, Vaccine 17 (5) (1999) 466–472.

[56] E. Handman, Leishmaniasis: current status of vaccine development, Clin. Microbiol. Rev. 14 (2) (2001) 229–243.

[57] F. Modabber, Vaccines against leishmaniasis, Ann. Trop. Med. Parasitol. 89 (Suppl. 1) (1995) 83–88.
[58] D.M. Russo, J.M.J. Burns, E.M. Carvalho, R.J. Armitage, K.H. Grabstein, L.L. Button, W. R. McMaster, S.G. Reed, Human T cell responses to gp63, a surface antigen of Leishmania, J. Immunol. 147 (1991).
[59] A. Sjölander, S. Ståhl, P. Perlmann, Bacterial expression systems based on protein A and protein G designed for the production of immunogens: applications to *Plasmodium falciparum* malaria antigens, ImmunoMethods 2 (1998) 79–92.
[60] J.M. Kelley, Genetic transformation of parasitic protozoa, Adv. Parasitol. 39 (1997) 227–270.
[61] R.A. Schwendener, Liposomes as vaccine delivery systems: a review of the recent advances, Ther. Adv. Vaccines 2 (6) (2014) 159–182.
[62] A. Kumar, M. Samant, DNA vaccine against visceral leishmaniasis: a promising approach for prevention and control, Parasite Immunol. 38 (5) (2016) 273–281.
[63] D. Rivier, P. Bovay, R. Shah, S. Didisheim, J. Mauel, Vaccination against *Leishmania major* in a CBA mouse model of infection: role of adjuvants and mechanism of protection, Parasite Immunol. 21 (1999) 461–473.
[64] S. Gurunathan, D.L. Sacks, D.R. Brown, S.L. Reiner, H. Charest, N. Glaichenhaus, et al., Vaccination with DNA encoding the immunodominate LACK parasite antigen confers protective immunity to mice infected with *Leishmania major*, J. Exp. Med. 186 (1997) 1137–1147.
[65] T. ten Broeke, R. Wubbolts, W. Stoorvogel, MHC class II antigen presentation by dendritic cells regulated through endosomal sorting, Cold Spring Harb. Perspect. Biol. 5 (2013), a016873.
[66] S.B. Ahmed, C. Bahloul, C. Robbana, S. Askri, K. Dellagi, A comparative evaluation of different DNA vaccine candidates against experimental murine leishmaniasis due to *L. major*, Vaccine 22 (13–14) (2004) 1631–1639.

Current status of *Leishmania* vaccine and problems to be solved

13

Current status of *Leishmania* vaccine

It is necessary to understand and summarize the vaccine status of all forms of leishmaniasis. Therefore, important vaccination approaches for overall leishmaniasis are summarized in Table 13.1. These were only studied previously to understand the overall picture/status of vaccination.

Problems to be solved before a *Leishmania* vaccine becomes reality

After a lot of work and effort done in Leishmania vaccine research (Chapter 12 and this chapter), there are no *Leishmania* vaccines approved by the US Food and Drug Administration (FDA) available on the market. Here, the author discusses some problems that might need to be hurdled in the development of a Leishmania vaccine. The following points or problems must be solved before a *Leishmania* vaccine becomes a reality.

a. Protective antigens

When complex organisms such as protozoan parasites interact with the immune system, the induction of antibodies or cell-mediated immune responses takes place because most of their components are immunogenic to the host, as shown in the literature. However, responses to some antigens have no value in terms of host protection. Some may even contribute to pathological responses by cross-reactivity with host molecules [110]. In the early days of vaccine development, when live or killed whole parasites were considered as vaccine candidates, the identification of antigens that induce the appropriate immune response leading to host protection was only a dream. The advent of gene cloning and monoclonal antibodies opened the possibility of identifying and characterizing relevant protein antigens and, most importantly, producing them in unlimited amounts. It is now clear that many antigens may be combined to elicit protective immune responses, and many more will be discovered when the sequencing of the *Leishmania* genome is completed. It is somewhat interesting that the same gene cloning technology that allowed the isolation of genes encoding vaccine candidates can now be used to engineer the ablation of essential genes in *Leishmania*, thus making the use of live attenuated vaccines a much more

Visceral Leishmaniasis. https://doi.org/10.1016/B978-0-323-91124-5.00007-2

Table 13.1 *Leishmania* vaccine agents used in different animal models and their success rate.

Vaccine	Agent	Animal Model	Success rate	References
Killed vaccine	*L. amazonesis* or mixer of five killed Leishmania stocks	Human	Significant protection from natural infection	[1]
Infection by *L. chagasi*	*L. chagasi*	Dogs	Possible etiopathogenetic mechanism	[2]
Killed vaccine	Combination of killed *L. mexicana* or *L. braziliensis* *M. bovis* BCG	Human	High cure rate even in patients with severe infection	[3] [4]
Live vaccine	Live promastigotes of *L. major*	Human	Not very successful	[5]
Killed vaccine	Mixed BCG—*L. major*	Human	No protection, transient stimulation	[6]
Killed vaccine	*L. amazonensis* with recombinant IL-12	Monkey, mice dogs	Human trials	[7]
Live attenuated vaccine	Irradiated promastigotes	Mice	Good	[8] [9]
Live attenuated promastigotes	*L. major*, *L. tropica*, *L. mexicana*	Mice	Good	[10] [11] [12]
Recombinant or native gp63 synthetic peptides	*L. amazonensis*	Mice, primates	Good	[13] [14] [15]
Recombinant or native gp46/M2/PSA-2	*L. major*, *L. mexicana*, *L. donovani*	Mice	Excellent but dependent on conformation adjuvant	[16] [17]
Recombinant LACK	Promastigotes, amastigotes	BALB/c mice	Good, enhanced by IL-12	[18]
A2, P4, P8	*L. mexicana*	Mice	Good	[19] [20]

Flagellar antigen LCR 1	*L. donovani* *L. chagasi*	Mice	Good	[21]
Nonprotein antigen	*Leishmania* Lipophospho glycon (LPG)		Excellent protection and dependent on the use of adjuvant	[22] [23]
Naked DNA gp63, PSA-2, and LACK	*L. major*	Mice	Good	[24] [25]
Subunit vaccine based on recombinant PSA-2, gp63	*L. major*	Mice	Variable to poor	[26]
Live virulent	*L. major*, *L. topica*	Human	Effective	[27] [5]
Live promastigotes	Zoonotic cutaneous leishmaniasis (*L. major*)	Human	93%	[28] [29] [30]
Living heterologous parasite	*L. tropica* and *L. mexicana*	Human	Positive skin test	[31] [32]
Living heterologous parasite	*L. enrietti*, *L. dononani*, *L. Mexicana*, and L. *brazielensis*	Guinea pigs	Good	[33] [34]
Living heterologous parasite	*L. major*, *L. enrietti*, *L. donovani*, and L. *tropica*	Mouse and human	Good	[35]
Living heterologous parasite	*L. major*, *L. donovani*	Vervet monkey	High degree of resistance	[14] [15]
Living heterologous parasite	*L. major*, *L. panamensis*	Hamster	Significant resistance	[36]
Living heterologous parasite (*L. brazilensis*)	*L. major*	BALB/c mice	Good	[37]
Living heterologous parasite (Dihydrofolate reductase thymidylate synthase-DHFR-TS of *L. major*)	*L. amazonensis*	BALB/c mice, C57BL/6 mice		[38]
Virulent parasite	*L. major*	BALB/c mice		[39] [40]
Natural avirulent parasite	*L. donovani*	Wild squirrel		[41] [42]

Continued

Table 13.1 Continued

Vaccine	Agent	Animal Model	Success rate	References
UR6 strain (*L. donovani*) subcutaneous	*L. donovani*	Mice	Protection found	[42]
UR6 promastigote (*L. donovani*) intravenous route	*L. donovani*	Mice	No protection found	[42]
Virulent promastigotes (subcutaneous)	*L. chagasi*	Murine model	Protection found	[43]
Cloned avirulent *L. major* intraperitoneally	*L. major*	Mice	Good	[44,45]
Irradiated amastigotes intraperitoneal with *L. donovani*	*L. donovani*	Hamster	Confers protection	[46]
Radio-attenuated amastigotes (*L. major)*	*L. mexicana*	CBA mice	Successful	[11]
γ-Irradiated promastigote surface protein gp63	*L. major*	CBA mice, BALB/c mice	Not satisfactory	[8]
DHFR-TS mutant of *L. major*	*L. major*	BALB/c mice	Satisfactory	[47] [48] [49]
DHFR-TS mutant of *L. major*	*L. major*, *L. amazonensis*	Human cells *in vitro*	Substantially protected	[38]
L. mexicana wild-type strain lacking CPA or CPB both	*L. mexicana*	BALB/c mice	Good	[11]
Inactivated BT1 by homologous recombination	WT parasite	BALB/c mice	Successful	[50]
DHFR-TS gene subcutaneous inoculation	Virulent *L. chagasi*, *L. major*, *L. donovani*	BALB/c mice	Satisfactory	[43]
Dead promastigote	*L. major*	BALB/c mice	Successful	[51]
ALM + BCG	*L. infantum*	Dogs	Successful, inducing immune response	[52]
Merthiolated *L. brazilensis* promastigote + BCG	*L. chagasi*	Dogs	Satisfactory	[53]

ALM + BCG	*L. infantum*	Dogs	Prospects for a canine Leishmania vaccine	[54]
Autoclaved freeze thawed *L. amazonensis* in alum + recombinant IL-12 as adjuvant	*L. amazonensis*	*Rhesus macaques* (Monkey)	Successful, inducing immune response	[7]
ALM + BCG with alum (subcutaneous)	*L. donovani*	Indian langur	Successful in Indian langur	[55]
Phenol killed promastigote *L. brazilensis*, *L. guyanensis*, *L. amazonensis*	*L. brazilensis*, *L. guyanensis*, *L. amazonensis*	Human	72.9%	[56]
Killed Leishmania promastigotes + BCG	*L. amazonensis*	Human	Satisfactory	[57]
ALM + BCG	CL (zoonotic)	Human	16.5%	[6]
BCG	CL	Human	3.2%	[6]
β-1,3-glucan with some fraction of *L. donovani*	*L. donovani*	Mice	Good	[58]
Microsomal fraction of *L. donovani*	*L. donovani*	Golden hamster	Protective effect	[59]
Inoculum with *Corneybacterium pavarum*	*L. major*	BALB/c mice	Protects mice	[60]
LACK	*L. major*	BALB/c mice		[61]
Live promastigote	*L. donovani*	Human	80%	[32]
Live promastigote	*L. donovani*	Human	No protection	[62]
10^7 live amastigotes	*L. donovani*	Hamster	Four-fold reduction	[63]
10^2 promastigote	*L. chagasi*	BALB/c mice	No protection	[43]
10^4 live promastigotes	*L. chagasi*	BALB/c mice	Intermediate protection	[43]
10^7 live promastigotes	*L. chagasi*	BALB/c mice	88%	[43]
10^7 live promastigotes	*L. major*	BALB/c mice	No protection	[43]
Attenuated 10^7 promastigotes	*L. chagasi*	BALB/c mice	No protection	[43]
10^7 pro. (DHFR-TS gene knockout)	*L. chagasi*	BALB/c mice	No protection	[43]
5×10^7 promastigote (i.v.) BT1 Null mutant	*L. donovani*	BALB/c	80% (L), 90% (S)	[50]
10^8 live parasites	*L. donovani* (UR6 strain)	(Hamster)	86% (L), 88% (S)	[42]

Continued

Table 13.1 Continued

Vaccine	Agent	Animal Model	Success rate	References
Live BCG	BCG	BALB/c	Partial protection	[64]
Live BCG	BCG	Hamster	Exacerbation	[65]
Adoptive transfer of immunized spleen cells	*L. donovani*	BALB/c	Partial	[66]
ALM + BCG	Autoclaved *L. major*	Canine	CMI response	[52]
ALM + BCG	Autoclaved *L. major*	Primate	Partial	[67]
ALM + BCG	Autoclaved *L. major*	Human	LST positive-52%	[68]
ALM + BCG	Autoclaved *L. major*	Human	LST positive-62%	[69]
Alum + ALM + BCG	Autoclaved *L. major*	Primate	95% cured	[55]
ALD + BCG	Autoclaved *L. donovani*	Hamster	86% (S)	[70]
10^8 formaldehyde killed promastigotes	*L. donovani* (UR6)	Hamster	95% (L), 94% (S)	[42]
10^8 heat-killed promastigotes	*L. donovani* (UR6)	Hamster	91% (L), 91% (s)	[42]
10^8 pro. Sonicated	*L. donovani* (UR6)	Hamster	99% (L), 95% (S)	[42]
10^8 heat-killed promastigote	*L. donovani*	BALB/c mice	80% (L)	[71]
AM antigen on NCP	AM antigen	C57BL/6 mice	Partial	[72]
LAg in positive liposome	LAg	BALB/c	87% (L), 81% (S)	[73]
LAg in positive liposome	LAg	Hamster	97% (S)	[73]
LAg in neutral liposome	LAg	BALB/c	73% (L), 69% (S)	[74]
LAg in negative liposome	LAg	BALB/c	59% (L), 57% (S)	[46]
LAg in DSPC liposome	LAg	BALB/c	93% (L), 96% (S)	[75]
Dendritic cells pulsed with SLA	SLA	BALB/c	10–1000 fold reduction	[76]

SLA + IL-12	SLA	BALB/c	50% (L)	[77]
Dp72 + *C. parvum*	Dp72	BALB/c	78% (L)	[78]
Aluminum hydroxide	FML	Swiss albino	68% (L)	[79]
FML + Aluminum hydroxide	FML	Swiss albino	88% (L)	[80]
FML + FIA	FML	Swiss albino	No protection	[81]
FML + BCG	FML	Swiss albino	52% (L)	[81]
FML + IL-12	FML	Swiss albino	Failure (L)	[81]
FML+ QS 21	FML	Swiss albino	79% (L)	[81]
FML + Quil A	FML	Swiss albino	79% (L)	[81]
FML+ Riedel De Haen	FML	Swiss albino	73% (L)	[81]
FML + Quil A	FML	Canine	95% protection	[82]
ORFF + CFA	LD1 complex	BALB/c	40%–50% (L) partial	[83]
BT1 + CFA	LD1 complex	BALB/c	40%–50% (L) partial	[83]
ORFF + BT1 + CFA	LD1 complex	BALB/c	70%–80% (L) partial	[83]
ORFF DNA	ORFF DNA	BALB/c	80% (L), 60% (S)	[84]
rHASPB1	rHASPB1	BALB/c	91% (L), 90% (S)	[77]
rHASPB1 + IL12	rHASPB1	BALB/c	78% (L), intermediate (S)	[77]
15 cDNA sublibraries	CDNA library	BALB/c	1000-fold protection	[85]
BCG expressing LCR1	LCR1	BALB/c	Partial	[21]
Only papLe 22 cDNA	*L. donovani*	Hamster	50% reduction in parasite circulation	[86]
Promastigote surface protease gp63 (leishmanolysin) + adjuvant	*L. major*	Mice	Protection	[87] [9]
Synthetic T cell epitope (subcutaneous) + poloxamer as adjuvant	*L. major* virulent	BALB/c mice	Long-term protection	[26]
Expression libraries of *L. major*	*L. major*	Mice	Satisfactory	[88]
Oligonucleotide primer from *L. donovani* DNA	*L. donovani*	Mouse	Induces response	[89]
Cysteine proteinases (cp) *{L.major}*	*L. major*	Mice	Good immune response	[90]

Continued

Table 13.1 Continued

Vaccine	Agent	Animal Model	Success rate	References
GRP-78 (*L. donovani* + Freunds complete adjuvant 10^5 live) *L. major*	*L. major*	C57BL/6 mice	Highly efficient	[91]
Histone H-1 (*L. major*) by perchloric extraction	*L. major*	BALB/c mice		[92]
gp-46 /M-2+ *C. pavarum* (*L. mexicaca*)	*L. Mexicana, L. amazonensis*	Mouse	Partial to complete protection	[16]
Gp63 (rgp63) rMBAP + recombinant cysteine proteinae5 (rCP5)	*L. mexicana*	C57BL/5 mice	Protective	[93]
L. pifanoi amastigote antigen P-8 + *C. pavarum*	*L. amazonensis* *L. pifanoi*	Mouse	Partial to complete protection	[94]
rA2 + heat-killed *P. acnes*	rA2	BALB/c	89% (L)	[19]
A2 DNA	rA2	BALB/c	65% (L)	[95]
A2 DNA + inhibition of p53 response	rA2	BALB/c	80% (L)	[95]
Gp36 + Reidel De Haen	Gp36	BALB/c	68% (L)	[96]
IL-12 p40 –p35 fusion cDNA	P36 (LACK DNA)	BALB/c	No protection	[97–99]
LACK DNA+ vaccinia virus	LACK DNA	Canine	60% protection	[100]
Eluted gp63 in positive liposome	gp63	BALB/c	57% (L) 60% (S)	[46]
Q protein formulated with BCG	Q protein	Canine	90% protection	[101]
Infection with *L. major* by bites of *Phlebotomus duboscqi*	*L. major*	Vervet monkey	Satisfactory	[102]
Mixture of TSA + LmSTI-1 + LeiF (Leish-IIIF)	*L. major*	Murine BALB/c primates	Immense protection	[103]

Recombinant LCR1	*L. chagasi*	BALB/c mice	Partial protection	[104]
L. infantum pro. Fraction (LiF2)	*L. major*	BALB/c C57BL/6 mice	Strong protection	[105]
FML of *L. donovani*	*L. donovani*	Dogs	Strong long-lasting protection	[106]
Subcutaneous inoculation of p-36 with intraperitoneal infection of vvp36 IL-12	*L. major*	BALB/c mice		[107]
Attenuated *Salmonella typhimurium* expressed gp-63 from *L. maxicana*	*L. maxicana*	BALB/c, XC57BL/6	Partial protection	[108]
Salmonella typhimurium expressed gp-63 from *L. major*	*L. major*	CBA mice		[109]

attractive proposition than ever before. The wheel seems to have turned full circle; we may not need to know which are the most highly protective antigens [111].

Data obtained using different strains of inbred mice indicate that genetic variation in the host has a major influence in determining the outcome of infection. Part of this variation probably reflects differences in the ability of the host to respond to individual antigens. A high responder to one antigen may be a low responder to another antigen. Accordingly, a vaccine to be used in an outbred population, such as humans, will probably require several different antigens to guarantee a satisfactory overall response by most if not all of the population. One could envisage the design of polyvalent vaccines incorporating as many polypeptide and glycolipid antigens as possible. While the large-scale production of polypeptides may not be a problem, the large-scale production of complex glycolipid antigens may be difficult. The production of strings of antigenic protein epitopes (known as polytopes) might be considered, as has been done for the Ebna virus [112]. Such combinations may be necessary to ensure that all individuals in the population, including low responders to some antigens, will get the benefit of a vaccine.

b. Safety concerns

One of the nightmares in vaccine development is the possibility that a particular vaccine may exacerbate the disease associated with the infection [113] or cause pathological reactions due to cross-reactivity between the host and parasite antigens. Another nightmare is the possibility of unleashing the disease due to incomplete attenuation or ineffective killing of the organisms. There is also the possibility that attenuated organisms may cause disease in immunocompromised individuals. In the current worldwide AIDS pandemic, such individuals are no longer rare [114]. People in tropical countries rarely suffer from a single disease and many have lower immunocompetence, as already demonstrated by low responses to various other vaccines. Many are malnourished or starving, a significant risk factor for visceral leishmaniasis (VL) [110].

c. Delivery and adjuvants

Adjuvants, from the Latin word meaning "to help," have been used to improve vaccine efficacy since the early 1920s [115]. While the number and type of adjuvants have expanded, their mechanism of action has remained largely mysterious and empirical. Recent studies have changed the thinking about cells and molecules that initiate immune responses. The current concept is that primary T-cell responses must be initiated by specialized "professional" antigen-presenting cells. These cells are the dendritic cells or the Langerhans cells in the skin [116]. The antigen is taken up by dendritic cells via receptor-mediated endocytosis or fluid-phase pinocytosis. The interaction with antigen drives the dendritic cells to mature and migrate out of the tissue into the draining lymph nodes. This is where the antigen is processed into peptides and where the peptides meet the MHC class II molecules as they are assembled and transported through the endoplasmic reticulum and the Golgi apparatus. The peptide MHC class II complex is transported to the cell surface and displayed for recognition by T cells. The antigen-presenting cells also secrete cytokines that attract $CD4^+$ T-cells to the area, such that T-cells with cognate receptors for the peptide MHC class

II complex undergo clonal expansion. It is likely that the Th1/Th2 switch will occur at this point, thus determining the course and outcome of the immune response. It is at this level that adjuvants are believed to contribute to the amplitude of the immune response and to its quality. This scenario underlines the importance of a judicious choice of adjuvant for vaccines that require Th1 responses for protection [111].

The ability of bacteria and other particles to be taken up by dendritic cells, coupled with the ability to express foreign genes in bacteria, has made them attractive delivery vehicles for vaccines. Such a vaccine could exploit attenuated bacteria such as *Salmonella* or BCG, which are already in use as vaccines with demonstrated safety and immunogenicity in their own right. BCG has been used successfully for anti-*Leishmania* immunotherapy in South American patients without side effects. BCG vectors carrying gp63 have also been used successfully to induce protection in the *L. major* system [93]. *S. enterica* serovar *Typhimurium* expressing gp63 has also been used in the mouse model. Unfortunately, the vaccine induced variable protection ranging from minimal to significant, despite the induction of apparently appropriate T-cell responses. It is possible that the use of "adjuvanted" bacteria carrying the genes for TNF-α and IFN-γ, which appear to potentiate the immune response, may improve the performance of a *Salmonella*-delivered vaccine [108]. An added advantage of a *Salmonella*-delivered vaccine is its oral administration. Problems with the *Salmonella* vectors include the instability of the plasmids carrying the genes of interest and the need for multiple administrations [108,109].

Several other types of particulate adjuvants or delivery systems have been tested for use with *Leishmania* vaccines, including include liposomes, microparticles, immunostimulating complexes, and micelles formed by intrinsically adjuvanted lipopeptides. The outcomes have been variable. Most of this work currently involves animal models, and the use of these systems in humans is still somewhat distant [111].

d. Use of cytokines as an adjuvant

One of the most promising adjuvants involves the use of some "natural adjuvant," such as soluble cytokines, which are known to promote Th1 immune responses. Among these, IL-12 is essential for the induction and maintenance of Th1 immune responses by *Leishmania* vaccines, including DNA vaccines [117]. It appears that the persistence of IL-12 delivered as DNA may be an important contributor to the long-term memory induced by the vaccinating DNA-encoded LACK antigen.

The timing of the administration of the cytokine in relation to the vaccine has a major effect on the quality and maintenance of the response. In experiments testing the ability of IL-12 to amplify the vaccinating effect of PSA-2 DNA, a paradoxical effect of IL-2 was observed. While mice injected with DNA encoding IL-12 alone showed protection from infection, protection was abrogated in mice injected with a combination of IL-12 and PSA-2-encoding DNA [111].

An interesting approach to the targeting of the immune response has been coupling the potent antigen-presenting ability of dendritic cells to the delivery of proinflammatory cytokines to the local site of response. For this purpose, the adoptive transfer of dendritic cells engineered by retroviral infection to secrete IL-12 has been used to augment the effect of vaccination with dendritic cells pulsed with *L. donovani* antigen. This regimen led to an increase in parasite-specific IFN-γ production and 1–3 log unit lower parasite burdens in mice with murine visceral leishmaniasis [76].

e. Immunostimulatory effects of bacterial DNA

Certain sequences of bacterial DNA seem to have immunostimulatory effects [118]. Specific DNA sequences containing unmethylated dinucleotides (CpG motifs) activate B cells and dendritic cells and induce cytokine production by macrophages. The ability of CpG motifs to induce the production of IL-12 and TNF-α and to lead to a polarized Th1 response makes them particularly attractive as adjuvants for *Leishmania* vaccines. CpG-containing DNA has immunostimulatory effects in vaccination against *L. major* [119,120]. DNA may become an important and relatively nontoxic adjuvant for *Leishmania* vaccines.

References

[1] R.X. Armijos, M.M. Weigel, M. Calvopina, A. Hidalgo, W. Cevallos, J. Correa, Safety, immunogenecity, and efficacy of an autoclaved *Leishmania amazonensis* vaccine plus BCG adjuvant against New World cutaneous leishmaniasis, Vaccine 22 (2004) 1320–1326.

[2] A.P. Giannuzzi, M. Ricciardi, A. De Simone, F. Gernone, Neurological manifestations in dogs naturally infected by *Leishmania infantum*: descriptions of 10 cases and a review of the literature, J. Small Anim. Pract. 58 (3) (2017) 125–138.

[3] M. Cabrera, J.M. Blackwell, M. Castes, D. Trujillo, J. Convit, M.A. Shaw, Immunotherapy with live BCG plus heat killed Leishmania induces a T helper 1-like response in American cutaneous leishmaniasis patients, Parasite Immunol. 22 (2000) 73–79.

[4] M. Castes, J. Blackwell, D. Trujillo, S. Formica, M. Cabrera, G. Zorrilla, A. Rodas, P.L. Castellanos, J. Convit, Immune response in healthy volunteers vaccinated with killed leishmanial promastigotes plus BCG. I: Skin-test reactivity, T-cell proliferation and interferon-gamma production, Vaccine 12 (1994) 1041–1051.

[5] C.L. Greenblatt, D.T. Spira, B. Montilio, H. Gerichter, An improved protocol for the preparation of a frozen promastigote vaccine for cutaneous leishmaniasis, J. Biol. Stand. 8 (1980) 227–232.

[6] I. Sharifi, A.R. FeKri, M.R. Aflatonian, A. Khamesipour, A. Nadim, M.R. Mousavi, A.Z. Momeni, Y. Dowlati, T. Godal, F. Zicker, P.G. Smith, F. Modabber, Randomised vaccine trial of single dose of killed *Leishmania major* plus BCG against anthroponotic cutaneous leishmaniasis in Bam, Iran, Lancet 351 (1998) 1540–1543.

[7] R.T. Kenney, D.L. Sacks, J.P. Sypek, L. Vilela, A.A. Gam, K. Evans-Davis, Protective immunity using recombinant human IL-12 and alum as adjuvants in a primate model of cutaneous leishmaniasis, J. Immunol. 163 (1999) 4481–4488.

[8] D. Rivier, R. Shah, P. Bovay, J. Mauel, Vaccine development against cutaneous leishmaniasis. Subcutaneous administration of radioattenuated parasites protects CBA mice against virulent *Leishmania major* challenge, Parasite Immunol. 15 (1993) 75–84.

[9] D. Rivier, P. Bovay, R. Shah, S. Didisheim, J. Mauel, Vaccination against *Leishmania major* in a CBA mouse model of infection: role of adjuvants and mechanism of protection, Parasite Immunol. 21 (1999) 461–473.

[10] J. Alexander, R.S. Phillips, *Leishmania tropica* and *Leishmania mexicana*: cross-immunity in mice, Exp. Parasitol. 45 (1978) 93–100.

[11] J. Alexander, A radioattenuated *Leishmania major* vaccine markedly increases the resistance of CBA mice to subsequent infection with *Leishmania mexicana* mexicana, Trans. R. Soc. Trop. Med. Hyg. 76 (1982) 646–649.
[12] D.G. Russell, The macrophage-attachment glycoprotein gp63 is the predominant C3-acceptor site on *Leishmania mexicana* promastigotes, Eur. J. Biochem. 164 (1987) 213–221.
[13] D.M. Russo, A. Jardim, E.M. Carvalho, P.R. Sleath, R.J. Armitage, R.W. Olafson, S.G. Reed, Mapping human T cell epitopes in Leishmania gp63: identification of cross-reactive and species-specific epitopes, J. Immunol. 150 (1993) 4713.
[14] M.M. Gicheru, J.O. Olobo, C.O. Anjili, Heterologous protection by *Leishmania donovani* for *Leishmania major* infections in the vervet monkey model of the disease, Exp. Parasitol. 85 (1997) 109–116.
[15] M.M. Gicheru, J.O. Olobo, C.O. Anjili, A.S. Orago, F. Modabber, P. Scott, Vervet monkeys vaccinated with killed *Leishmania major* parasites and interleukin-12 develop a type 1 immune response but are not protected against challenge infection, Infect. Immun. 69 (2001) 245–251.
[16] J. Champsi, D. McMahon-Pratt, Membrane glycoprotein M-2 protects against *Leishmania amazonensis* infection, Infect. Immun. 56 (1988) 3272–3279.
[17] E. Handman, F.M. Symons, T.M. Baldwin, J.M. Curtis, J.P. Scheerlinck, Protective vaccination with promastigote surface antigen 2 from *Leishmania major* is mediated by a TH1 type of immune response, Infect. Immun. 63 (1995) 4261–4267.
[18] S. Gurunathan, D.L. Sacks, D.R. Brown, S.L. Reiner, H. Charest, N. Glaichenhaus, R.A. Seder, Vaccination with DNA encoding the immunodominant LACK parasite antigen confers protective immunity to mice infected with *Leishmania major*, J. Exp. Med. 186 (1997) 1137–1147.
[19] A. Ghosh, W.W. Zhang, G. Matlashewski, Immunization with A2 protein results in a mixed Th1/Th2 and a humoral response which protects mice against *Leishmania donovani* infections, Vaccine 20 (2001) 59–66.
[20] S. Kar, L. Soong, M. Colmenares, K. Goldsmith-Pestana, D. McMahon-Pratt, The immunologically protective P-4 antigen of Leishmania amastigotes. A developmentally regulated single strand-specific nuclease associated with the endoplasmic reticulum, J. Biol. Chem. 275 (2000) 37789–37797.
[21] J.A. Streit, T.J. Recker, J.E. Donelson, M.E. Wilson, BCG expressing LCR1 of *Leishmania chagasi* induces protective immunity in susceptible mice, Exp. Parasitol. 94 (2000) 33–41.
[22] E. Handman, H.M. Jarvis, Nitrocellulose-based assays for the detection of glycolipids and other antigens: mechanism of binding to nitrocellulose, J. Immunol. Methods 83 (1985) 113–123.
[23] M.J. McConville, A. Bacic, G.F. Mitchell, E. Handman, Lipophosphoglycan of Leishmania major that vaccinates against cutaneous leishmaniasis contains an alkyl glycerophosphoinositol lipid anchor, Proc. Natl. Acad. Sci. U. S. A. 84 (1987) 8941–8945.
[24] A. Sjolander, T.M. Baldwin, J.M. Curtis, E. Handman, Induction of a Th1 immune response and simultaneous lack of activation of a Th2 response are required for generation of immunity to leishmaniasis, J. Immunol. 160 (1998) 3949–3957.
[25] P.S. Walker, T. Scharton-Kersten, E.D. Rowton, U. Hengge, A. Bouloc, M.C. Udey, J.C. Vogel, Genetic immunization with glycoprotein 63 cDNA results in a helper T cell type 1 immune response and protection in a murine model of leishmaniasis, Hum. Gene Ther. 9 (1998) 1899–1907.

[26] N. Spitzer, A. Jardim, D. Lippert, R.W. Olafson, Long-term protection of mice against *Leishmania major* with a synthetic peptide vaccine, Vaccine 17 (1999) 1298–1300.
[27] S. Frankenburg, O. Axelrod, S. Kutner, C.L. Greenblatt, S.N. Klaus, E.A. Pirak, R. McMaster, G.H. Lowell, Effective immunization of mice against cutaneous leishmaniasis using an intrinsically adjuvanted synthetic lipopeptide vaccine, Vaccine 14 (1996) 923–929.
[28] A. Nadim, E. Javadian, G. Tahvildar-Bidruni, M. Ghorbani, Effectiveness of leishmanization in the control of cutaneous leishmaniasis, Bull. Soc. Pathol. Exot. Filiales 76 (1983) 377–383.
[29] F. Modabber, Experiences with vaccines against cutaneous leishmaniasis: of men and mice, Parasitology 98 (Suppl) (1989) S49–S60.
[30] W. Peters, D.A. Evans, S.M. Lanham, Importance of parasite identification in cases of leishmaniasis, J. R. Soc. Med. 76 (1983) 540–542.
[31] S. Adler, A.E. Gunders, Immunity to *Leishmania mexicana* following spontaneous recovery from oriental sore, Trans. R. Soc. Trop. Med. Hyg. 58 (1964) 274–277.
[32] P.E. Manson-Bahr, R.B. Heisch, Transient infection of man with a Leishmania (*L. adleri*) of lizards, Ann. Trop. Med. Parasitol. 55 (1961) 381–382.
[33] R.A. Neal, R.A. Miles, The Montenegro reaction in Guinea pigs infected by *Leishmania enriettii* and the effect of antigens prepared from various Leishmania isolates, Am. J. Trop. Med. Hyg. 79 (1976) 32–37.
[34] A.D. Bryceson, R.S. Bray, R.A. Wolstencroft, D.C. Dumonde, Immunity in cutaneous leishmaniasis of the Guinea-pig, Clin. Exp. Immunol. 7 (3) (1970 Sep) 301–341.
[35] D.J. Wyler, F.I. Weinbaum, H.R. Herrod, Characterization of in vitro proliferative responses of human lymphocytes to leishmanial antigens, J. Infect. Dis. 140 (1979) 215–221.
[36] Y. Osorio, S.J. Gonzalez, V.L. Gama, B.L. Travi, Reinfection in American cutaneous Leishmaniasis: evaluation of clinical outcomes in the hamster model, Mem. Inst. Oswaldo Cruz 93 (3) (1998) 353–356.
[37] H.C. Lima, G.K. DeKrey, R.G. Titus, Resolution of an infection with *Leishmania braziliensis* confers complete protection to a subsequent challenge with *Leishmania major* in BALB/c mice, Mem. Inst. Oswaldo Cruz 94 (1999) 71–76.
[38] P. Veras, C. Brodskyn, F. Balestieri, L. Freitas, A. Ramos, A. Queiroz, A. Barral, S. Beverley, M. Barral-Netto, A dhfr-ts-Leishmania major knockout mutant cross-protects against *Leishmania amazonensis*, Mem. Inst. Oswaldo Cruz 94 (1999) 491–496.
[39] J. Dietrich, C. Aagaard, R. Leah, A.W. Olsen, A. Stryhn, T.M. Doherty, et al., Exchanging ESAT6 with TB10.4 in an Ag85B fusion molecule-based tuberculosis subunit vaccine: efficient protection and ESAT6-based sensitive monitoring of vaccine efficacy, J. Immunol. 174 (2005) 6332–6339.
[40] S. Kamhawi, The biological and immunomodulatory properties of sand fly saliva and its role in the establishment of Leishmania infections, Microbes Infect. 2 (2000) 1765–1773.
[41] D. Heyneman, Immunology of leishmaniasis, Bull. World Health Organ. 44 (4) (1971) 499–514.
[42] S. Mukhopadhyay, S. Bhattacharyya, R. Majhi, T. De, K. Naskar, S. Majumdar, S. Roy, Use of an attenuated leishmanial parasite as an immunoprophylactic and immunotherapeutic agent against murine visceral Leishmaniasis, Clin. Diagn. Lab. Immunol. 7 (2000) 233–240.
[43] J.A. Streit, T.J. Recker, F.G. Filho, S.M. Beverley, M.E. Wilson, Protective immunity against the protozoan *Leishmania chagasi* is induced by subclinical cutaneous infection with virulent but not avirulent organisms, J. Immunol. 166 (3) (2001) 1921–1929.

[44] G.F. Mitchell, E. Handman, T.W. Spithill, Vaccination against cutaneous leishmaniasis in mice using nonpathogenic cloned promastigotes of *Leishmania major* and importance of route of injection, Aust. J. Exp. Biol. Med. Sci. 62 (Pt 2) (1984) 145–153.
[45] E. Handman, R.E. Hocking, G.F. Mitchell, T.W. Spithill, Isolation and characterization of infective and non-infective clones of *Leishmania tropica*, Mol. Biochem. Parasitol. 7 (1983) 111–126.
[46] F. Afrin, R. Rajesh, K. Anam, M. Gopinath, S. Pal, N. Ali, Characterization of *Leishmania donovani* antigens encapsulated in liposomes that induce protective immunity in BALB/c mice, Infect. Immun. 70 (12) (2002) 6697–6706.
[47] R.G. Titus, J.M. Ribeiro, Salivary gland lysates from the sand fly Lutzomyia longipalpis enhance Leishmania infectivity, Science 239 (1988) 1306–1308.
[48] R.G. Titus, F.J. Gueiros-Filho, L.A. de Freitas, S.M. Beverley, Development of a safe live Leishmania vaccine line by gene replacement, Proc. Natl. Acad. Sci. U. S. A. 92 (1995) 10267–10271.
[49] R.G. Titus, C.M. Theodos, A.H. Shankar, L.R. Hall, Interactions between *Leishmania major* and macrophages, Immunol. Ser. 60 (1994) 437–459.
[50] B. Papadopoulou, G. Roy, M. Breton, C. Kundig, C. Dumas, I. Fillion, A.K. Singh, M. Olivier, M. Ouellette, Reduced infectivity of a *Leishmania donovani* biopterin transporter genetic mutant and its use as an attenuated strain for vaccination, Infect. Immun. 70 (2002) 62–68.
[51] J.G. Howard, F.Y. Liew, Mechanisms of acquired immunity in leishmaniasis, Philos. Trans. R. Soc. Lond. Ser. B Biol. Sci. 307 (1984) 87–98.
[52] S. Lasri, H. Sahibi, A. Sadak, C.L. Jaffe, A. Rhalem, Immune responses in vaccinated dogs with autoclaved *Leishmania major* promastigotes, Vet. Res. 30 (5) (1999) 441–449.
[53] W. Mayrink, O. Genaro, J.C. Silva, R.T. da Costa, W.L. Tafuri, V.P. Toledo, et al., Phase I and II open clinical trials of a vaccine against *Leishmania chagasi* infections in dogs, Mem. Inst. Oswaldo Cruz 91 (1996) 695–697.
[54] L. Gradoni, An update on antileishmanial vaccine candidates and prospects for a canine Leishmania vaccine, Vet. Parasitol. 100 (2001) 87–103.
[55] A. Misra, A. Dube, B. Srivastava, P. Sharma, J.K. Srivastava, J.C. Katiyar, S. Naik, Successful vaccination against *Leishmania donovani* infection in Indian langur using alum-precipitated autoclaved *Leishmania major* with BCG, Vaccine 19 (2001) 3485–3492.
[56] A. Giudice, I. Camada, P.T. Leopoldo, et al., Resistance of *Leishmania (Leishmania) amazonensis* and *Leishmania (Viannia) braziliensis* to nitric oxide correlates with disease severity in Tegumentary Leishmaniasis, BMC Infect Dis. 7 (2007) 7, https://doi.org/10.1186/1471-2334-7-7.
[57] F.Q. Grenfell Rafaella, A. Marques-da-Silva Eduardo, C. Souza-Testasicca Miriam, A.F. Coelho Eduardo, F.A. Paula, C. Afonso Luis Carlos, et al., Antigenic extracts of *Leishmania braziliensis* and *Leishmania amazonensis* associated with saponin partially protects BALB/c mice against *Leishmania chagasi* infection by suppressing IL-10 and IL-4 production, Mem. Inst. Oswaldo Cruz 105 (6) (2010) 818–822.
[58] J.A. Cook, T.W. Holbrook, Immunogenicity of soluble and particulate antigens from *Leishmania donovani*: effect of glucan as an adjuvant, Infect. Immun. 40 (1983) 1038–1043.
[59] K.A. Obaid, S. Ahmad, H.M. Khan, A.A. Mahdi, R. Khanna, Protective effect of *L. donovani* antigens using glucan as an adjuvant, Int. J. Immunopharmacol. 11 (3) (1989) 229–235.
[60] P. Scott, E. Pearce, S. Heath, A. Sher, Identification of T-cell-reactive antigens that protect BALB/c mice against *Leishmania major*, Ann. Inst. Pasteur. Immunol. 138 (1987) 771–774.

[61] A. Gumy, J.A. Louis, P. Launois, The murine model of infection with *Leishmania major* and its importance for the deciphering of mechanisms underlying differences in Th cell differentiation in mice from different genetic backgrounds, Int. J. Parasitol. 34 (2004) 433–444.
[62] S. Pampiglione, P.E. Manson-Bahr, M. La Placa, M.A. Borgatti, S. Musumeci, Studies in Mediterranean leishmaniasis. 3. The leishmanin skin test in kala-azar, Trans. R. Soc. Trop. Med. Hyg. 69 (1975) 60–68.
[63] H.R. Rezai, J. Farrell, E.L. Soulsby, Immunological responses of *L. donovani* infection in mice and significance of T cell in resistance to experimental leishmaniasis, Clin. Exp. Immunol. 40 (1980) 508–514.
[64] L.L. Smrkovski, C.L. Larson, Effect of treatment with BCG on the course of visceral leishmaniasis in BALB/c mice, Infect. Immun. 16 (1977) 249–257.
[65] J.C. Jarecki-Black, A.B. Glassman, T.W. Holbrook, The effect of BCG-vaccine upon experimental visceral leishmaniasis in hamsters, Ann. Clin. Lab. Sci. 14 (1984) 464–466.
[66] J.C. Jarecki-Black, A.B. Glassman, E.R. James, Adoptive transfer of vaccine-induced resistance to *Leishmania donovani*, Am. J. Trop. Med. Hyg. 34 (1985) 1095–1097.
[67] V.A. Dennis, W.L. Chapman Jr., W.L. Hanson, R. Lujan, Leishmania donovani: clinical, hematologic and hepatic changes in squirrel monkeys (*Saimiri sciureus*), J. Parasitol. 71 (1985) 576–582.
[68] E.A. Khalil, A.M. El Hassan, E.E. Zijlstra, M.M. Mukhtar, H.W. Ghalib, B. Musa, M.E. Ibrahim, A.A. Kamil, M. Elsheikh, A. Babiker, F. Modabber, Autoclaved Leishmania major vaccine for prevention of visceral leishmaniasis: a randomised, double-blind, BCG-controlled trial in Sudan, Lancet 356 (2000) 1565–1569.
[69] I.N. Satti, H.Y. Osman, N.S. Daifalla, S.A. Younis, E.A. Khalil, E.E. Zijlstra, A.M. El Hassan, H.W. Ghalib, Immunogenicity and safety of autoclaved *Leishmania major* plus BCG vaccine in healthy Sudanese volunteers, Vaccine 19 (2001) 2100–2106.
[70] J.K. Srivastava, A. Misra, P. Sharma, B. Srivastava, S. Naik, A. Dube, Prophylactic potential of autoclaved *Leishmania donovani* with BCG against experimental visceral leishmaniasis, Parasitology 127 (2003) 107–114.
[71] M.A. Bebars, A.O. el Serougi, K.M. Makled, E.M. Mikhael, M.M. Abou Gamra, M. el Sherbiny, A.W. Mohareb, E.A. Mohammed, An experimental vaccine providing heterologous protection for Leishmania species in murine model, J. Egypt. Soc. Parasitol. 30 (2000) 137–156.
[72] P.M. Kaye, A.J. Curry, J.M. Blackwell, Differential production of Th1- and Th2-derived cytokines does not determine the genetically controlled or vaccine-induced rate of cure in murine visceral leishmaniasis, J. Immunol. 146 (1991) 2763–2770.
[73] F. Afrin, N. Ali, Adjuvanticity and protective immunity elicited by *Leishmania donovani* antigens encapsulated in positively charged liposomes, Infect. Immun. 65 (1997) 2371–2377.
[74] N. Ali, F. Afrin, Protection of mice against visceral leishmaniasis by immunization with promastigote antigen incorporated in liposomes, J. Parasitol. 83 (1997) 70–75.
[75] A. Askarizadeh, M.R. Jaafari, A. Khamesipour, A. Badiee, Liposomal adjuvant development for leishmaniasis vaccines, Ther. Adv. Vaccines 5 (4–5) (2017) 85–101.
[76] S.S. Ahuja, R.L. Reddick, N. Sato, E. Montalbo, V. Kostecki, W. Zhao, M.J. Dolan, P.C. Melby, S.K. Ahuja, Dendritic cell (DC)-based anti-infective strategies: DCs engineered to secrete IL-12 are a potent vaccine in a murine model of an intracellular infection, J. Immunol. 163 (1999) 3890–3897.
[77] S. Stager, D.F. Smith, P.M. Kaye, Immunization with a recombinant stage-regulated surface protein from *Leishmania donovani* induces protection against visceral leishmaniasis, J. Immunol. 165 (2000) 7064–7071.

[78] N. Rachamim, C.L. Jaffe, Pure protein from *Leishmania donovani* protects mice against both cutaneous and visceral leishmaniasis, J. Immunol. 150 (1993) 2322–2331.
[79] W.R. Santos, R.R. Bernardo, L.M. Pecanha, M. Palatnik, J.P. Parente, C.B. Palatnik de Sousa, Haemolytic activities of plant saponins and adjuvants. Effect of Periandra mediterranea saponin on the humoral response to the FML antigen of *Leishmania donovani*, Vaccine 15 (1997) 1024–1029.
[80] W.R. Santos, E. Paraguai de Souza, M. Palatnik, C.B. Palatnik de Sousa, Vaccination of Swiss Albino mice against experimental visceral leishmaniasis with the FML antigen of *Leishmania donovani*, Vaccine 17 (20–21) (1999) 2554–2561.
[81] W.R. Santos, V.M. de Lima, E.P. de Souza, R.R. Bernardo, M. Palatnik, C.B. Palatnik de Sousa, Saponins, IL12 and BCG adjuvant in the FML-vaccine formulation against murine visceral leishmaniasis, Vaccine 21 (2002) 30–43.
[82] G.P. Borja-Cabrera, A. Cruz Mendes, E. Paraguai de Souza, L.Y. Hashimoto Okada, F.A. Trivellato, J.K. Kawasaki, A.C. Costa, A.B. Reis, O. Genaro, L.M. Batista, M. Palatnik, C.B. Palatnik-de-Sousa, Effective immunotherapy against canine visceral leishmaniasis with the FML-vaccine, Vaccine 22 (2004) 2234–2243.
[83] V.S. Dole, V.S. Raj, A. Ghosh, R. Madhubala, P.J. Myler, K.D. Stuart, Immunization with recombinant LD1 antigens protects against experimental leishmaniasis, Vaccine 19 (4–5) (2000) 423–430.
[84] B. Sukumaran, P. Tewary, S. Saxena, R. Madhubala, Vaccination with DNA encoding ORFF antigen confers protective immunity in mice infected with *Leishmania donovani*, Vaccine 21 (2003) 1292–1299.
[85] P.C. Melby, G.B. Ogden, H.A. Flores, W. Zhao, C. Geldmacher, N.M. Biediger, S.K. Ahuja, J. Uranga, M. Melendez, Identification of vaccine candidates for experimental visceral leishmaniasis by immunization with sequential fractions of a cDNA expression library, Infect. Immun. 68 (2000) 5595–5602.
[86] K. Fragaki, I. Suffia, B. Ferrua, D. Rousseau, Y. Le Fichoux, J. Kubar, Immunisation with DNA encoding *Leishmania infantum* protein papLe22 decreases the frequency of parasitemic episodes in infected hamsters, Vaccine 19 (2001) 1701–1709.
[87] B. Taillan, P. Marty, S. Schneider, H. Telle, J.G. Fuzibet, E. Rosenthal, A. Rahal, Y. Lefichoux, P. Dujardin, Visceral leishmaniasis involving a cutaneous Kaposi's sarcoma lesion and free areas of skin, Eur. J. Med. 1 (1992) 255.
[88] D. Piedrafita, et al., Regulation of macrophage IL-12 synthesis by Leishmania phosphoglycans, Eur. J. Immunol. 29 (1999) 235–244.
[89] A. Campos-Neto, L. Soong, J.L. Cordova, D. Sant'Angelo, Y.A. Skeiky, N.H. Ruddle, S. G. Reed, C. Janeway Jr., D. McMahon-Pratt, Cloning and expression of a *Leishmania donovani* gene instructed by a peptide isolated from major histocompatibility complex class II molecules of infected macrophages, J. Exp. Med. 182 (1995) 1423–1433.
[90] S. Rafati, A.H. Salmanian, K. Hashemi, C. Schaff, S. Belli, N. Fasel, Identification of *Leishmania major* cysteine proteinases as targets of the immune response in humans, Mol. Biochem. Parasitol. 113 (2001) 35–43.
[91] A.T. Jensen, J. Curtis, J. Montgomery, E. Handman, T.G. Theander, Molecular and immunological characterisation of the glucose regulated protein 78 of Leishmania donovani(1), Biochim. Biophys. Acta 1549 (2001) 73–87.
[92] N. Solioz, U. Blum-Tirouvanziam, R. Jacquet, et al., The protective capacities of histone H1 against experimental murine cutaneous leishmaniasis, Vaccine 18 (9–10) (1999) 850–859.
[93] T. Aebischer, M. Wolfram, S.I. Patzer, T. Ilg, M. Wiese, P. Overath, Subunit vaccination of mice against new world cutaneous leishmaniasis: comparison of three proteins expressed in amastigotes and six adjuvants, Infect. Immun. 68 (2000) 1328–1336.

[94] L. Soong, S.M. Duboise, P. Kima, D. McMahon-Pratt, Leishmania pifanoi amastigote antigens protect mice against cutaneous leishmaniasis, Infect. Immun. 63 (9) (1995) 3559–3566, https://doi.org/10.1128/IAI.63.9.3559-3566.1995.
[95] A. Ghosh, S. Labrecque, G. Matlashewski, Protection against *Leishmania donovani* infection by DNA vaccination: increased DNA vaccination efficiency through inhibiting the cellular p53 response, Vaccine 19 (2001) 3169–3178.
[96] I. Aguilar-Be, R. da Silva Zardo, E. Paraguai de Souza, et al., Cross-protective efficacy of a prophylactic *Leishmania donovani* DNA vaccine against visceral and cutaneous murine leishmaniasis, Infect. Immun. 73 (2) (2005) 812–819.
[97] P.C. Melby, B. Chandrasekar, W. Zhao, J.E. Coe, The hamster as a model of human visceral leishmaniasis: progressive disease and impaired generation of nitric oxide in the face of a prominent Th1-like cytokine response, J. Immunol. 166 (2001) 1912–1920.
[98] P.C. Melby, A. Tabares, B.I. Restrepo, A.E. Cardona, H.S. McGuff, J.M. Teale, Leishmania donovani: evolution and architecture of the splenic cellular immune response related to control of infection, Exp. Parasitol. 99 (2001) 17–25.
[99] P.C. Melby, J. Yang, W. Zhao, L.E. Perez, J. Cheng, Leishmania donovani p36(LACK) DNA vaccine is highly immunogenic but not protective against experimental visceral leishmaniasis, Infect. Immun. 69 (2001) 4719–4725.
[100] M.J. Ramiro, J.J. Zarate, T. Hanke, D. Rodriguez, J.R. Rodriguez, M. Esteban, J. Lucientes, J.A. Castillo, V. Larraga, Protection in dogs against visceral leishmaniasis caused by *Leishmania infantum* is achieved by immunization with a heterologous prime-boost regime using DNA and vaccinia recombinant vectors expressing LACK, Vaccine 21 (2003) 2474–2484.
[101] I. Molano, M.G. Alonso, C. Miron, E. Redondo, J.M. Requena, M. Soto, C.G. Nieto, C. Alonso, A *Leishmania infantum* multi-component antigenic protein mixed with live BCG confers protection to dogs experimentally infected with *L. infantum*, Vet. Immunol. Immunopathol. 92 (2003) 1–13.
[102] A.A. Binhazim, J.I. Githure, G.K. Muchemi, G.D. Reid, Isolation of *Leishmania major* from a naturally infected vervet monkey (*Cercopithecus aethiops*) caught in Kiambu District, Kenya, J. Parasitol. 73 (6) (1987) 1278–1279.
[103] A. Campos-Neto, J.R. Webb, K. Greeson, R.N. Coler, Y.A. Skeiky, S.G. Reed, Vaccination with plasmid DNA encoding TSA/LmSTI1 leishmanial fusion proteins confers protection against *Leishmania major* infection in susceptible BALB/c mice, Infect. Immun. 70 (6) (2002) 2828–2836.
[104] M.E. Wilson, B.M. Young, K.P. Andersen, J.V. Weinstock, A. Metwali, K.M. Ali, J.E. Donelson, A recombinant *Leishmania chagasi* antigen that stimulates cellular immune responses in infected mice, Infect. Immun. 63 (1995) 2062–2069.
[105] L. Monjour, I. Vouldoukis, B.W. Ogunkolade, C. Hetzel, M. Ichen, D. Frommel, Vaccination and treatment trials against murine leishmaniasis with semi-purified Leishmania antigens, Trans. R. Soc. Trop. Med. Hyg. 82 (1988) 412–415.
[106] N. Santarem, A. Tomas, A. Ouaissi, J. Tavares, N. Ferreira, A. Manso, L. Campino, J.M. Correia, A. Cordeiro-da-Silva, Antibodies against a *Leishmania infantum* peroxiredoxin as a possible marker for diagnosis of visceral leishmaniasis and for monitoring the efficacy of treatment, Immunol. Lett. 101 (2005) 18–23.
[107] E. Tapia, E. Perez-Jimenez, L. Lopez-Fuertes, R. Gonzalo, M.M. Gherardi, M. Esteban, The combination of DNA vectors expressing IL-12 + IL-18 elicits high protective immune response against cutaneous leishmaniasis after priming with DNA-p36/LACK and the cytokines, followed by a booster with a vaccinia virus recombinant expressing p36/LACK, Microbes Infect. 5 (2003) 73–84.

[108] D. Xu, S.J. McSorley, S.N. Chatfield, G. Dougan, F.Y. Liew, Protection against *Leishmania major* infection in genetically susceptible BALB/c mice by GP63 delivered orally in attenuated *Salmonella typhimurium* ($AroA^-$ $AroD^-$), Immunology 85 (1995) 1–7.
[109] D.M. Yang, N. Fairweather, L.L. Button, W.R. McMaster, L.P. Kahl, F.Y. Liew, Oral *Salmonella typhimurium* (AroA-) vaccine expressing a major leishmanial surface protein (gp63) preferentially induces T helper 1 cells and protective immunity against leishmaniasis, J. Immunol. 145 (7) (1990) 2281–2285.
[110] E. Handman, Leishmania virulence: it's a knock out! Trends Parasitol. 17 (2001) 60.
[111] E. Handman, Leishmaniasis: current status of vaccine development, Clin. Microbiol. Rev. 14 (2001) 229–243.
[112] W. Zeng, D.C. Jackson, J. Murray, K. Rose, L.E. Brown, Totally synthetic lipid-containing polyoxime peptide constructs are potent immunogens, Vaccine 18 (2000) 1031–1039.
[113] T.W. Hoskins, J.R. Davies, A.J. Smith, C.L. Miller, A. Allchin, Assessment of inactivated influenza-A vaccine after three outbreaks of influenza A at Christ's Hospital, Lancet i (1979) 33–35.
[114] F.J. Wright, Exacerbation of dimorphous leprosy apparently due to BCG vaccination, Lepr. Rev. 42 (1971) 219–221.
[115] V.E.J.C. Schijns, Immunological concepts of vaccine adjuvant activity, Curr. Opin. Immunol. 12 (2000) 456–463.
[116] J. Banchereau, R.M. Steinman, Dendritic cells and the control of immunity, Nature 392 (1998) 245–252.
[117] P. Scott, G. Trinchieri, IL-12 as an adjuvant for cell-mediated immunity, Semin. Immunol. 9 (1997) 285–291.
[118] S. Yamamoto, E. Kuramoto, S. Shimada, T. Tokunaga, In vitro augmentation of natural killer cell activity and production of interferon-alpha/beta and -gamma with deoxyribonucleic acid fraction from *Mycobacterium bovis* BCG, Jpn. J. Cancer Res. 79 (1988) 866–873.
[119] P.S. Walker, T. Scharton-Kersten, A.M. Krieg, L. Love-Homan, E.D. Rowton, M.C. Udey, J.C. Vogel, Immunostimulatory oligodeoxynucleotides promote protective immunity and provide systemic therapy for leishmaniasis via IL-12- and IFN-gamma-dependent mechanisms, Proc. Natl. Acad. Sci. U. S. A. 96 (1999) 6970–6975.
[120] S. Zimmermann, O. Egeter, S. Hausmann, G.B. Lipford, M. Rocken, H. Wagner, K. Heeg, CpG oligodeoxynucleotides trigger protective and curative Th1 responses in lethal murine leishmaniasis, J. Immunol. 160 (1998) 3627–3630.

Amastigote-based vaccine: Silver lining in leishmaniasis

14

Amastigote-based vaccine

Most studies related to the *Leishmania* vaccine are done with the promastigote antigen [1,2]. However, some studies have been done on amastigote antigens (Table 14.1) and they were found to provide more significant protection against infection. This, however, could be explored more in-depth. Table 14.1 summarizes some significant studies that have been done via selecting amastigote antigens for vaccine development.

Amastigote-specific antigens have been considered promising candidates for the prevention of leishmaniasis because amastigote stage of parasites persist infections in the host. Antigens selecting from the amastigote stage could be considered to improve the level of vaccine-induced cell-mediated immunity and protection against leishmaniasis. Ufm1, BT1, the A2 gene, P27, centrin, etc. [16] could be important amastigote-based vaccine candidates that could be important in a future Leishmania vaccine.

Visceral Leishmaniasis. https://doi.org/10.1016/B978-0-323-91124-5.00003-5

Table 14.1 Previous studies done by selecting amastigote antigens for *Leishmania* vaccine development.

***Leishmania* species**	**Forms of Leishmaniasis**	**Antigen isolate form**	**Remarks**	**References**
Leishmania donovani	Visceral	Amastigote-specific hypothetical protein called LiHyJ	Recombinant protein plus Th1 adjuvant or DNA plasmid-based vaccine	[3]
Leishmania infantum	Visceral	Leishmania amastigote 2 (A2) protein	Formulations to be translated into human clinical trials	[4]
Leishmania infantum	Visceral	*Leishmania* amastigote-specific protein (LiHyp1)	Vaccine for canine visceral leishmaniasis	[5]
A good vaccine candidate for L. *chagasi* and L. *amazonensis* vaccine	Visceral and cutaneous	Protein or DNA vaccine in a murine model of VL	A2 antigen of amastigote was tested as a vaccine candidate	[6]
L. major	Cutaneous	Amastigote class I nuclease (LmaCIN) is highly expressed in the amastigote stage of L. *major*	Homologous to the P4 nuclease of *L. pifanoi*, and cured human patients	[7]
L. pifanoi (Axenic amastigotes)	Cutaneous	Antigen P-2	Amastigote induces MHC II-dependent NKC reactivity in cells from donors	[8]
L. donovani (Axenic amastigotes)	Visceral	Recombinant A2 antigen	Diagnosis of American leishmaniasis in humans and dogs	[9]
L. donovani (Axenic amastigotes)	Visceral	Recombinant A2 antigen	Mixed Th1/Th2 and a humoral response in mice BALB/c (Lshs, H-2d) and C57B/6 mice	[10]
L. donovani (Axenic amastigotes)	Visceral leishmaniasis	200-kDa amastigote-specific antigen	Evaluation of *L. donovani* by enzyme-linked immunosorbent assay (ELISA)	[11]

L. amazonensis (Amastigote of cutaneous lesions)	Cutaneous	Antigen-30 kDa-termed p30	Murine model	[12]
Leishmania donovani amastigotes	Visceral	Isolation of lipophosphoglycans	Lipophosphoglycan molecules, structurally distinct from the previously characterized LPG of the promastigote stage	[13]
L. pifanoi (Axenic amastigotes)	Cutaneous	Antigens, P-2, P-4, and P-8	Antigens protect against *L. amazonensis* in mice	[14]
L. amazonensis (Amastigote isolate)	Cutaneous	Identification of glycosphingolipid antigen of amastigote	Cutaneous lesions of hamster	[15]

References

[1] M. Colmenares, M. Tiemeyer, P. Kima, D. McMahon-Pratt, Biochemical and biological characterization of the protective Leishmania pifanoi amastigote antigen P-8, Infect. Immun. 69 (11) (2001) 6776–6784.

[2] J. Sadlova, J. Myskova, T. Lestinova, J. Votypka, M. Yeo, P. Volf, Leishmania donovani development in Phlebotomus argentipes: comparison of promastigote- and amastigote-initiated infections, Parasitology 144 (4) (2017) 403–410.

[3] J.A. Oliveira-da-Silva, A.S. Machado, F.F. Ramos, G.S.V. Tavares, D.P. Lage, D.V.C. Mendonça, I.A.G. Pereira, T.T.O. Santos, V.T. Martins, L.M. Carvalho, C.S. Freitas, F. Ludolf, T.A.R. Reis, R.S. Bandeira, A.M. Silva, L.E. Costa, J.S. Oliveira, M.C. Duarte, B.M. Roatt, A.L. Teixeira, E.A.F. Coelho, A *Leishmania* amastigote-specific hypothetical protein evaluated as recombinant protein plus Th1 adjuvant or DNA plasmid-based vaccine to protect against visceral leishmaniasis, Cell Immunol. 356 (2020) 104194.

[4] A.P.M.M. Almeida, L.F.M. Machado, D. Doro, F.C. Nascimento, L. Damasceno, R.T. Gazzinelli, A.P. Fernandes, C. Junqueira, New vaccine formulations containing a modified version of the amastigote 2 antigen and the non-virulent *Trypanosoma cruzi* CL-14 strain are highly antigenic and protective against *Leishmania infantum* challenge, Front. Immunol. 9 (2018) 465.

[5] V.T. Martins, M.A. Chávez-Fumagalli, L.E. Costa, A.M.C. Canavaci, P.S. Lage, et al., Correction: antigenicity and protective efficacy of a Leishmania amastigote-specific protein, member of the super-oxygenase family, against visceral Leishmaniasis, PLoS Negl. Trop. Dis. 7 (4) (2013), https://doi.org/10.1371/annotation/427655ac-c278-41f5-95ca-6279c562752f.

[6] F.H. Zanin, E.A. Coelho, C.A. Tavares, E.A. Marques-da-Silva, M.M. Silva Costa, S.A. Rezende, R.T. Gazzinelli, A.P. Fernandes, Evaluation of immune responses and protection induced by A2 and nucleoside hydrolase (NH) DNA vaccines against *Leishmania chagasi* and *Leishmania amazonensis* experimental infections, Microbes Infect. 9 (2007) 1070–1077.

[7] S. Farajnia, F. Mahboudi, S. Ajdari, N.E. Reiner, A. Kariminia, M.H. Alimohammadian, Mononuclear cells from patients recovered from cutaneous leishmaniasis respond to *Leishmania major* amastigote class I nuclease with a predominant Th1-like response, Clin. Exp. Immunol. 139 (3) (2005 Mar) 498–505.

[8] S. Nylén, K. Maasho, D. McMahon-Pratt, H. Akuffo, Leishmanial amastigote antigen P-2 induces major histocompatibility complex class II-dependent natural killer-cell reactivity in cells from healthy donors, Scand. J. Immunol. 59 (3) (2004) 294–304.

[9] F.A. Carvalho, H. Charest, C.A. Tavares, G. Matlashewski, E.P. Valente, A. Rabello, R.T. Gazzinelli, A.P. Fernandes, Diagnosis of American visceral leishmaniasis in humans and dogs using the recombinant *Leishmania donovani* A2 antigen, Diagn. Microbiol. Infect. Dis. 43 (4) (2002 Aug) 289–295.

[10] A. Ghosh, W.W. Zhang, G. Matlashewski, Immunization with A2 protein results in a mixed Th1/Th2 and a humoral response which protects mice against *Leishmania donovani* infections, Vaccine 20 (1–2) (2001) 59–66.

[11] P. Kaul, N. Malla, S. Kaur, R.C. Mahajan, N.K. Ganguly, Evaluation of a 200-kDa amastigote-specific antigen of *L. donovani* by enzyme-linked immunosorbent assay (ELISA) for the diagnosis of visceral leishmaniasis, Trans. R. Soc. Trop. Med. Hyg. 94 (2) (2000 Mar-Apr) 173–175.

[12] C.G. Beyrodt, A.R. Pinto, E. Freymüller, C.L. Barbiéri, Characterization of an antigen from *Leishmania amazonensis* amastigotes able to elicit protective responses in a murine model, Infect. Immun. 65 (6) (1997) 2052–2059.

[13] T.B. McNeely, P.S. Doyle, Isolation of lipophosphoglycans from *Leishmania donovani* amastigotes, Arch. Biochem. Biophys. 334 (1) (1996) 1–8.

[14] L. Soong, S.M. Duboise, P. Kima, D. McMahon-Pratt, Leishmania pifanoi amastigote antigens protect mice against cutaneous leishmaniasis, Infect. Immun. 63 (1995) 3559–3566.

[15] E.M. Saraiva, P.F. Pimenta, M.E. Pereira, W. de Souza, Isolation and purification of amastigotes of *Leishmania mexicana* amazonensis by a gradient of Metrizamide, J. Parasitol. 69 (3) (1983) 627–629.

[16] S.C. Pandey, A. Kumar, M. Samant, Genetically modified live attenuated vaccine: a potential strategy to combat visceral leishmaniasis, Parasite Immunol. 42 (9) (2020) e12732.

Index

Note: Page numbers followed by *f* indicate figures and *t* indicate tables.

Kolibri 360 GmbH
Ferdinand-Jühlke-Straße 7
99095 Erfurt, Deutschland
produktsicherheit@kolibri360.de